Poisons of the Past

Family of peasants in seventeenth-century France.
(*Le Retour du baptême* by the Le Nain brothers.)
From the Musée du Louvre, Paris. Reprinted with permission
of the Cliché des Musées Nationaux, Paris.

Poisons of the Past

Molds, Epidemics, and History

Mary Kilbourne Matossian

Yale University Press

New Haven and London

Acknowledgment is made for permission to include material based on the following: for chapter 2, "Climate, diet, and population growth in rural Russia, 1860–1913," *Slavic Review* 45, no. 3 (1986), pp. 457–469; for chapter 6, "The time of the Great Fear," *Sciences* (March–April 1984), pp. 38–41; for chapter 7, "Mold poisoning and population growth in England and France, 1750–1850," *Journal of Economic History* 44 (1984), pp. 669–686; for chapter 8, "The throat distemper reappraised," *Bulletin of the History of Medicine* 54 (1980), pp. 529–543; and for chapter 9, "Ergot and the Salem witchcraft affair," *American Scientist* (July–August 1982), pp. 185–192.

Designed by James J. Johnson
and set in Trump Roman types by
Brevis Press, Bethany, Connecticut.
Printed in the United States of America by
BookCrafters, Chelsea, Michigan.

Library of Congress Cataloging-in-Publication Data

Matossian, Mary Allerton Kilbourne.
   Poisons of the past: molds, epidemics, and history / Mary
Kilbourne Matossian.
      p. cm.
   Bibliography: p.
   Includes index.
   ISBN 0–300–03949–2 (alk. paper)
      1. Mycotoxicoses—Social aspects—Europe—History.   2. Food
poisoning—Social aspects—Europe—History.   3. Mycotoxicoses—
Social aspects—New England—History.   4. Food poisoning—Social
aspects—New England—History.   5. Diseases and history—Europe.
6. Diseases and history—New England.
RA1242.M94M38   1989
614.4   89–5345
                                                                         CIP

The paper in this book meets the guidelines for
permanence and durability of the Committee on
Production Guidelines for Book Longevity of the
Council on Library Resources.

3 5 7 9 10 8 6 4

Contents

List of Figures   vii

List of Tables   ix

Preface   xi

I. INTRODUCTION

1. Food Poisoning and History   3
2. A Case Study: Russia and Its Neighbors   22

II. CONTRIBUTIONS TO A HEALTH HISTORY OF EUROPE

3. A New Look in the Distant Mirror   47
4. Mycotoxins and Health in Early Modern Europe   59
5. Witch Persecution in Early Modern Europe   70
6. The Great Fear of 1789   81
7. The Population Explosion of 1750–1850   88

III. CONTRIBUTIONS TO A HEALTH HISTORY
OF COLONIAL NEW ENGLAND

8. The Throat Distemper   107
9. Ergot and the Salem Witchcraft Affair   113
10. Great Awakening or Great Sickening?   123

IV. REFLECTIONS

11. Social Control of Mass Psychosis   145
12. Plant Health and Human Health   155

Notes   159

Index   185

Figures

Frontispiece. Family of peasants in seventeenth-century France

 1. Rye and wheat infected with ergot   8
 2. Life cycle of ergot and the risks of mycotoxicosis   10
 3. Mean crude death rates in European Russia, 1861–1913   31
 4. Mean infant mortality rates in European Russia,
    1867–1910   32
 5. Infant mortality and January temperatures, 1861–1913   33
 6. Mean crude birth rates in European Russia, 1861–1913   35
 7. Alkaloid yields of *Claviceps* strains in Russian provinces   40
 8. Index of population in eastern Normandy, 1250–1550   54
 9. Index of population in Périgueux, 1330–1490   54
10. Distribution of rye cultivation in England and Wales   62
11. Distribution of witch trials in western Europe,
    1580–1650   72
12. Distribution of witch trials in Scotland   73
13. Distribution of episodes of panic, 1789, and production
    of rye and maslin in France   85
14. Production of four grains in East Worcestershire,
    1540–1867   93
15. Sum of specific death ratios of the eight leading
    causes of death in London, 1750–1909   97
16. Percent of deaths from "convulsions" in London,
    1750–1909   98
17. Percent of deaths from "consumption" in London,
    1750–1909   99
18. Percent of population growth in different regions of
    France, 1700–1787   100

19. Summer temperatures in North America and Europe   108
20. Location of dwellings of the bewitched in Salem Village   121
21. Topography of the main part of Connecticut   134
22. The negative supernatural interpretation of
    nervous disorders   146
23. The positive supernatural interpretation of
    nervous disorders   148

Tables

1. Symptoms of ergotism   11
2. Symptoms of alimentary toxic aleikiia   16
3. Means of variables in Russia, 1885–1909   37
4. Means of variables in Europe, 1751–1880   41
5. Correlations between variables in Europe and Russia   42
6. Public health in northwestern Europe and Italy,
   1351–1499   53
7. Correlations of crude birth rate and crude
   death rate with economic and temperature
   variables in England, 1660–1739   67
8. Population growth in selected countries of Europe,
   1750–1850   88
9. Mortality of English peers' children, 1550–1849   95
10. Tree-ring indices in Nancy Brook, New Hampshire,
    1650–1759   120
11. Admissions to the Congregational churches of
    Connecticut, 1735–1743   135

Preface

People make history, but not just as they please. However strong and intelligent they may be, human beings are vulnerable creatures. They may be overrun by foreign invaders, eclipsed by upstart political rivals, or their skills are made redundant by technologies emerging in other parts of the world. They may even be laid low by organisms too small to be seen by the naked eye.

Of these microorganisms those that cause infectious diseases are familiar to the historian. No study of Europe in the Middle Ages can ignore the effects of the bacterial infection known as bubonic plague. Less well known are the microfungi that produce poisons in food. Aided by favorable climatic conditions, these organisms have poisoned the supply of staple cereals upon which human beings depend, causing widespread death, especially among infants and children. Toxicity threatens not only the physical well-being of those affected, however; the poisons of fungi, or mycotoxins, may alter the behavior of those who have ingested them. They have, in fact, caused mass psychosis in communities, regions, and even nations.

Microfungi compete with human beings for the supply of available food plants. Some infect grain growing in the field; others, harvested grain in storage. Food poisoning from mycotoxins was a more serious problem in the past than it is now. Mycotoxins limited human fertility. They were a cause of sudden death. By acting as immunosuppressants, damaging the body's powers to resist infection, they made populations all the more susceptible to infectious diseases. And, of equal importance to the historian, they induced hallucinations or feelings of numbness or suffocation—symptoms which in past eras might have been interpreted as bewitchment or possession by the devil.

The fact that mycotoxins were most often found in cereals is important. In previous centuries people consumed a higher proportion of cereals overall than they do today. Bread was called the staff of life: the poor ate two to three pounds of it a day, and not much else. If contaminated, therefore, the staff of life could be the scepter of death.

One purpose of this book is to show how mycotoxins reduced fertility and increased mortality by insidious poisoning of the food supply. People did not know what the poisons were or how they worked. Today, analyzing the evidence of health history, we are able to include among the explanations of population patterns the effects of food poisoning. Another purpose of this book is to show how mycotoxins may have caused mass hallucinations, witch persecutions, and panics. Neither the participants nor the observers experiencing such events could fully comprehend the happenings around them, but now it may be possible to come to a better understanding of historical examples of seemingly bizarre behavior.

But how will we know that any given population or community is affected by food poisoning or, more specifically, mycotoxins? Here we are aided by the history of health in Russia, where serious food poisoning problems continued into the twentieth century. The government brought them under control only after 1945. The reasons for this delay are complex and interesting, but they are not the main concern of this book. Rather, suffice it to note that the existence of the delay enables us to learn about conditions that existed in the West before 1750 but were poorly recorded and measured.

In most of Europe north of the Alps and Pyrenees, as in Russia, the staple food of the masses in the eighteenth century was bread made from rye—a food source not only for man but also for the ergot fungus. Dietary change—the substitution of potatoes for rye bread and the increased availability of white wheat bread—began to accelerate in western Europe around 1750, although the time of this change varied among regions. In Russia, however, the common folk could not afford to eat wheat bread and resisted the introduction of the potato. Homemade rye bread, often full of ergot and other dangerous mycotoxins, remained the staple food. Even late in the nineteenth century the diet of most Russians resembled that of western Europeans before 1750.

For the scholar this is fortunate. Between 1865 and 1914 the Russian Imperial Government collected valuable statistics on fertility and mortality within its realm. It set up a network of weather stations and kept track of which crops were being harvested in each province. Although the records of the causes of death were meager, records that provided indirect evidence of food poisoning were abundant and generally accurate. Consequently, the intelligent observers in tsarist Russia unwittingly created for the modern historian a window into the past, the past not merely of Russia but of all of Europe north of the Alps and Pyrenees.

Once evidence of food poisoning is found, it informs our view of population patterns. Outbreaks of ergotism, known as "the holy fire" and "St. Vitus's dance," could decimate a community as surely as could the Black Death. But ergot toxins were also responsible for bizarre behavior. The symptoms of panic, "religious" excitement, and bewitchment may have had a common biochemical origin, although the culturally mandated interpretations of such behavior were different in each case.

Such claims may seem excessive; however, I have written this book with a keen sense of limitation, knowing that I must approach a number of problems in a new way. When the territory is strange and the way uncharted, each explorer must create his or her own map. At the same time, one cannot help being aware of one's own ignorance.

Traditional history tends to rely on a lot of information about a very few people; I have made judgments on the basis of a little information about a lot of people. I like to think I speak in the spirit of science, since I use the logic that many scientists use. Speaking with such a spirit I may often suggest, but can rarely insist; may often ask, but can rarely answer; may measure the probable, but never approach certainty. Boldness is needed to tackle an important problem, enthusiasm to persevere in the search for a solution; but caution and modesty are needed to evaluate the results.

I am indebted to the University of Maryland for providing me with a summer grant (1982) to read in the Wellcome Library on the History of Medicine in London and the Bibliothèque Nationale in Paris. I am also grateful to the National Research Council and the Academy of Sciences of the U.S.S.R. for enabling me (in 1984) to use the great libraries of Moscow and Leningrad: the State Central Scientific Medical Library, the Lenin Library, the Library of the Historical Museum, the Saltykov-Shchedrin Library, and the Library of the Academy of Sciences.

Special thanks are due to the staff of the History of Medicine Division of the National Library of Medicine, Bethesda, Maryland, and most especially to James Cassedy and Dorothy Hanks. I owe much to Jayne McLean as well as to the reference libraries of the National Agricultural Library in Beltsville, Maryland; the National Oceanic and Atmospheric Administration Library in Rockville, Maryland; the Library of Congress; and the McKeldin Library, University of Maryland at College Park. My friends, the late Helmut Landsberg, John Duffy, William McNeill, John Post, and William T. Stead, M.D., have not only helped me by reviewing manuscripts, but have given me the moral support needed to bring this project to completion.

Poisons of the Past

PART I

Introduction

CHAPTER ONE

Food Poisoning and History

Historical demographers have shown that long life and good health are recent blessings even in advanced industrial countries. Before 1750 the life expectancy at birth of a member of the British peerage was only 36.7 years; a hundred years later it had risen to 58.4 years. Conditions were worse, and improvement slower, among the common folk of Europe, but even the poor in most western European countries were better off in 1850 than in 1750. Moreover, between 1750 and 1850 the population of Europe almost doubled.

Before 1750 good health was a privilege of wealth, and not even all among the rich enjoyed it. Underweight, stunted, sickly, and occasionally deranged, most of the common folk could not even dream of the feelings of well-being that many today claim as a right. People constantly struggled with Death, their efforts tending to end in a draw. After a mortality crisis more people in a community would marry, marry younger, and give birth to more children; in a normal year most communities managed to produce more newborns than corpses. But then another visitation of Death would wipe out these meager gains.

In explaining this cycle of good years and bad, population historians once thought in Malthusian terms. Many perceived a human population as self-regulating: increase in the death rate was a result of population pressure on scarce resources, for example, and fluctuations in the birth rate were attuned to the rising and falling availability of resources. This position has lost popularity because the statistical evidence for it is weak.

More recently some population historians, notably those asso-

3

ciated with Cambridge University, have suggested that increasing fertility in England was more important than declining mortality in causing rapid population growth in the period 1750–1850. They believe that the increasing availability of industrial jobs stimulated the rise in fertility. This position is controversial.[1] But other aspects of the problem are less disputed. In Europe generally, population growth occurred in the late eighteenth and nineteenth centuries even though fertility was declining.[2] Industrialization may ultimately be found to have caused population growth in England, but it cannot explain such growth elsewhere in Europe.

Ronald Lee demonstrated in 1973 a positive correlation between fertility and real wages in Britain between 1250 and 1700, but he found mortality was determined by unknown "exogenous" factors. Since fertility remained fairly constant, change in mortality was the chief determinant of population trends. The mortality rate influenced population size, which in turn had an effect on real wages.[3] In 1985 Lee argued that increasing population size "reflected" the increasing demand for labor, but that independent variation in birth and death rates, rather than the demand for labor, drove the oscillations in population over the long run.[4]

What were the unknown factors that controlled birth and death rates prior to 1750? No one seems very certain about the answer. Fernand Braudel affirmed that the changing food supply was critically important in the sixteenth to eighteenth centuries. "Peasants and crops," he said, "in other words food supplies and the size of population, these determined the destiny of the age."[5] No one would deny that food supply is important, but recent research has shown that the causal links between nutrition on the one hand and both fertility and mortality on the other are far from established. There is evidence that only when people are extremely malnourished or thin does fertility decline. Nor is mortality from famine a major restraint on population growth. In a recent article Susan Watkins and Etienne van der Walle concluded:

> The slow rate of population growth in the past, in the context of a world in which the ability to expand production changed only slowly, has tempted historians to interpret this long-run immobility as the result of constraints imposed by a necessary equilibrium between population and resources. . . . Dramatic mortality from famine, however, appears to have been relatively rare, and because populations could recuperate rather rapidly

from a crisis it appears that famine was not a major mechanism by which the balance between population and resources was maintained in the past.[6]

If not famine, what other causes of death could significantly affect the mortality rate? Infectious diseases must also be considered. Prior to the eighteenth century, however, records of cause of death are so sketchy that ex post facto diagnosis from reported symptoms only is usually difficult. The epidemiology of different fatal conditions is often more valuable in categorizing mortality.

One clue is provided by the date or month of death, for certain kinds of illness tend to prevail at different times of the year. Prior to 1750 north of the Alps and Pyrenees mortality peaked in spring and again in August or September in crisis years. For those assuming that infectious diseases were the most important cause of death, this bi-modal pattern is very puzzling. Infectious diseases peak in summer (typhoid, dysentery) or late fall or winter (diphtheria, scarlatina, meningitis, influenza, strep throat, pneumonia, typhus, and smallpox). Only deaths from the various forms of tuberculosis peak in the spring.

Another clue to the solution of the problem is that prior to 1750 Death preferred the company of cool, wet weather during the growing season and harvest. It often attacked after a severe winter. What are believed to be the sickliest centuries—the fifteenth and seventeenth—were cold and wet. This is puzzling because at the present time the death rate from infectious diseases is low if summers are wet, high if they are dry. Rainfall now has no influence on the incidence of respiratory diseases in any season.[7]

A third clue to the determination of cause of death in history, once infectious diseases are set aside, may be found in diet. There are three dimensions of good human nutrition: first, sufficient calories to meet energy requirements; second, the supply of a range of nutrients—proteins, vitamins, and minerals—needed to maintain health; and third, the absence of harmful substances in food.

No one has shown that a dramatic increase either in caloric intake or in the supply of essential nutrients occurred in the diet of most Europeans between 1750 and 1850.[8] We must therefore consider the possibility that the proportion of harmful substances in food declined in this period, and to do this we must examine the European diet.

The great masses of common folk in Europe depended upon cereals for at least a majority of their calories. During the eighteenth century French adult peasants ate two to three pounds of dark bread a day, when they could afford it. The rich and affluent, less than 5 percent of the population, preferred white wheat bread. It was not so much the color of one's skin as the color of one's bread that proclaimed social status. In the Mediterranean Basin the diet of the poor consisted of barley, buckwheat, wheat, and (after the sixteenth century) maize. North of the Alps and Pyrenees the poor depended on bread made of rye or a mixture of rye and other grains, barley, oats, and buckwheat.

All these staple starches come from plants that are seldom free of mold and other fungal contamination, even before harvesting. Some molds infect the growing plant; others flourish in stored cereals; still others develop in grain left in the fields over the winter.

Plant diseases, like the growth of plants themselves, have a marked seasonality. Moreover, fungi, which cause a large proportion of plant diseases, flourish in cool, damp weather. Given the importance of cereal in the diet in the past, what effect would contamination of this food source have on birth and death rates and, more generally, the health of the population?

Fungi, which feed on plants, are plants themselves, although they do not have chlorophyll as most higher plants do. Mushrooms, mildews, and molds are the most familiar kinds of fungus.

Many molds and other fungi produce natural chemicals that are toxic, or poisonous, to people. This has been true since the emergence of the human species. Fortunately, human beings have developed natural aversions to a large proportion of dangerous natural poisons: contaminated food seems "bad" either to the eye, nose, or taste buds. But some naturally occurring mold toxins, especially those of the *Fusarium* genus, are insidious: their presence in cereals is not obvious. *Fusarium* molds infecting wheat, rye, rice, barley, oats, and other grains have produced epidemics of poisoning in Russia, and they are suspected of causing a high incidence of esophageal cancer in Lixian, China.[9] (Not all chemicals produced by fungi are poisonous. Some, such as those produced in certain species of mushroom, are merely intoxicating; others—penicillin for example—are quite beneficial to man.)

Today grain that is threshed in the field, bulk dried, and stored in large silos is most vulnerable to mold attack. A majority of molds grow between 15° and 30° C, but the most dangerous molds thrive at lower temperatures. Water is most often the limiting factor in their growth. No mold can grow in stored grain with a water content of

13 percent or less. This is why the drying of cereals before storage is very important. Wet spring and summer weather favors the infection of growing crops in the field and causes a high moisture level in harvested crops. In western and central Europe freshly harvested cereals frequently have a moisture level of 16–18 percent; in European Russia, 15–20 percent; but in the midwestern United States it is often close to 14 percent.[10]

As can plainly be seen, a mold is a superficial growth, often woolly in texture, but fungi can take a variety of shapes. A fungus of the genus *Claviceps*, during part of its life cycle, has a hard, seedlike form, known as the sclerotium. The name for this group of fungus, *ergot*, is taken from the French word for "cockspur," which the dark sclerotium resembles. (See Figures 1 and 2.) Sclerotia store reserve food material (as a seed does), among which are four kinds of alkaloids. (Alkaloids are nitrogenous organic compounds found in plants.) Ingestion of these compounds can affect every system of the human body—the nervous system as well as the cardiovascular system. Sclerotia grow in the body of grasses, especially rye, and if present when the grain is harvested contaminate the food supply of humans and animals. Bread containing 2 percent ergot can cause a community-wide epidemic of ergotism.

The study of diseases caused by fungal poisoning, or mycotoxicoses, is a relatively new field, not included in the curriculum of most medical schools. As C. A. Linsell said in 1977 (and it is still true in 1989), the only references to mycotoxicoses in medical textbooks are confined to ergotism and mushroom poisoning. The standard textbooks on mycotoxins provide rich chemical and mycological data but little clinical information. Most physicians, therefore, are not well prepared to recognize mycotoxins as possible causal agents in disease.[11] They are, of course, quite able to evaluate relevant information about mycotoxins if they are made aware of their importance.

Mycotoxicology is also a new area of interest in microbiology. According to Pat B. Hamilton,

> It is arguable that mycotoxins offer a new challenge to the logical framework on which microbiology is erected. The time-honored concepts of cause and effect are not as clear and simple with mycotoxins as we would wish and frequently assume. We have nondescript diseases with multiple nondescript causes. . . . Nevertheless mycotoxicology reflects reality with all its excitement and complexity as do few other areas of science.[12]

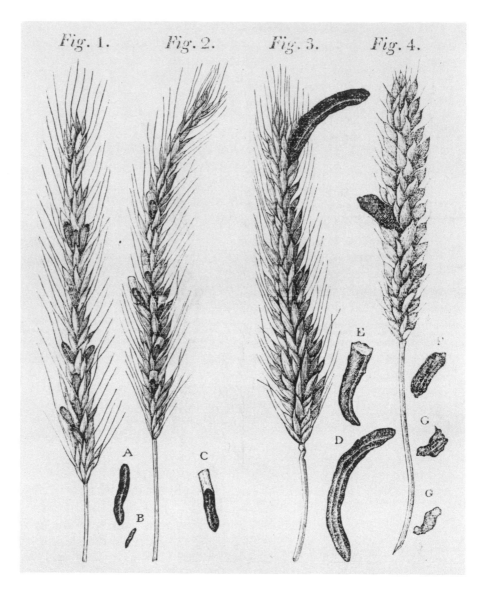

FIGURE I. Rye and wheat infected with ergot. The two ears of rye on the left contain many ergots, the third ear (of stout rye) only one large one. The ear on the right is wheat, only rarely infected with the fungus. Reprinted from Oliver Prescott, *A dissertation on the natural history and medicinal effects of the Secale cornutum, or ergot* (Boston, 1813), p. 12.

The commonest form of fungal poisoning, ergotism, has been known to educated people since the end of the eighteenth century, however.

ERGOT AND ERGOTISM

When ingested in sufficient quantity, ergot produces a disease called "ergotism" which has, in serious cases, two variants, convulsive and gangrenous ergotism.[13]

Victims of gangrenous ergotism may lose fingers, toes, and limbs to dry gangrene, caused by a vasoconstrictive chemical (such as the alkaloid ergotamine) produced by the ergot fungus.

Convulsive ergotism might better be labeled "dystonic ergotism." It is characterized by nervous dysfunction, such as writhing, tremors, and wry neck, which in the past were frequently reported as "convulsions" or "fits." It is now known that ergot alkaloids do not produce true convulsions, in which consciousness is lost, but some ergot alkaloids interfere with the activity of dopamine in the body, causing muscle spasms as well as confusion, delusions, and hallucinations. (See Table 1 for a list of the symptoms of ergotism.)

Various chemicals which constitute ergot interfere with the reproduction of humans and domestic animals. Ergonovine may act as an abortifacient, and other alkaloids suppress fertility or stop lactation. Some ergot toxins can pass through mother's milk and poison the nursing infant, who is especially vulnerable to such toxins. If lactating domestic animals in a community are also affected, there may be no alternative source of nutrition to human infants. The cumulative effect of these toxicoses might reduce reproductive levels.

Ergot might also produce temporary or permanent psychosis.[14] Among the known ergot alkaloids with hallucinogenic properties are ergine, ergonovine, and lysergic acid hydroxyethylamide.[15] In the laboratory it is now relatively easy to extract lysergic acid diethylamide (LSD) from lysergic acid, which is the basic ergot alkaloid. Moreover, through the action of other fungi, LSD may appear in natural ergot as well.[16]

N. N. Reformatskii made a careful study of mental disturbances associated with an epidemic of ergotism in Viatka province in northern Russia during 1889–1890.[17] In Nolinskii district he found that 30.1 percent of all persons hospitalized for ergotism suffered from mental disturbances. He believed that the percentage was higher

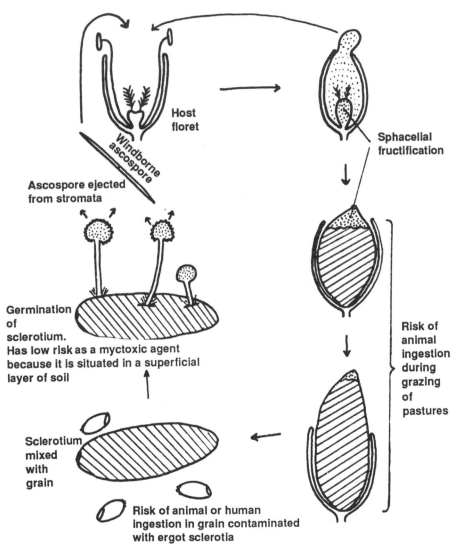

Conidia from sphacelial stage borne in honeydew
secretion of host plant; effect epidemic phase of disease

Host
floret

Windborne
ascospore

Sphacelial
fructification

Ascospore ejected
from stromata

Germination
of
sclerotium.
Has low risk as a myctoxic agent
because it is situated in a superficial
layer of soil

Risk of
animal
ingestion
during
grazing
of
pastures

Sclerotium
mixed
with
grain

Risk of animal or human
ingestion in grain contaminated
with ergot sclerotia

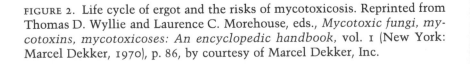

FIGURE 2. Life cycle of ergot and the risks of mycotoxicosis. Reprinted from Thomas D. Wyllie and Laurence C. Morehouse, eds., *Mycotoxic fungi, mycotoxins, mycotoxicoses: An encyclopedic handbook,* vol. 1 (New York: Marcel Dekker, 1970), p. 86, by courtesy of Marcel Dekker, Inc.

among those who were ill but not hospitalized. Unpleasant psychotic experiences were more common than pleasant ones. The unpleasant experiences included attacks of excitement and fear of enemies caused by visual illusions, hallucinations, and bad dreams.

Of the 52 cases of ergotism involving mental disturbances studied by Reformatskii, only 20 (38.1 percent) had gastrointestinal symptoms (nausea, diarrhea, and the like). As for age distribution, in Viatka province generally children accounted for only 18.1 percent of the sick but 40.9 percent of the fatal cases. Reformatskii studied

TABLE 1. Symptoms of ergotism

Cardiovascular system
   constriction of arteries and veins
   rapid, weak pulse
   precordial distress or pain
   muscle pain
   skin cold
   weakness, lameness
   gangrene
   cardiac arrest

Gastrointestinal system
   nausea
   vomiting
   diarrhea

Motor control
   tremors, spasms, writhing
   wry neck
   eyes awry
   loss of speech
   muscular paralysis
   renal spasm
   permanent constrictures

Central nervous system
   headache
   dizziness
   depression
   confusion
   drowsiness
   unconsciousness
   panic
   hallucinations
   delusions
   psychosis

Senses
 unquenchable thirst
 depressed or ravenous appetite
 sensations of heat ("fever") or cold ("chills")
 blindness
 deafness
 numbness
 feeling of being pinched, choked, suffocated

Skin
 tingling and itching ("formication")
 jaundice
 redness
 swelling
 blistering

Reproductive system
 fertility suppression
 abortion, stillbirth
 agalactia (inability to nurse)
 poisoning of mother's milk

hospitalized cases, and parents frequently did not hospitalize their sick children.

Speaking of ergotism epidemics in Russia generally, G. A. Kolosov reported that adolescents afflicted by the disease were especially prone to mental disorders.[18] If communities convinced that witches were in their midst were in fact suffering from ergotism epidemics, this may be why teenagers were so often the plaintiffs in witchcraft cases.

Unlike disease germs, poisons do not leave their victims with immunity upon recovery; indeed, for reasons not well understood, former sufferers of ergotism are more prone to the toxin afterwards. Infants and teenagers, too, are especially vulnerable because they consume more food per unit of body weight than adults; when that food is poisoned, they take in more poison per unit of body weight than adults.

Moreover, ergotism is a very lethal disease. During the ten epidemics recorded in Russia from 1832 to 1864, from 11 percent to 66 percent of those who took sick died, with a mean mortality rate of 41.5 percent.[19]

In the past ergotism was often mistaken for an infectious and contagious disease. Like such diseases it preferred the young as its victims, and it tended to affect many or all members of the same household more or less simultaneously. Victims of ergotism, like victims of infectious diseases, suffered from fever, and, as is true for victims of malaria, the fever might be intermittent, alternating with chills.

This complex of symptoms was sometimes labeled "ague," which researchers today may misinterpret to mean malaria. But ergotism was a killer disease; malaria, in its European forms, tended to weaken its victims without killing them. Furthermore, malaria first appeared in spring, ergotism in late summer.

The mold *Aspergillus fumigatus*, common in stored cereals, produces metabolites that can cause tremors and convulsions in men and animals. Some investigators think it can produce hallucinations as well, for some of its metabolites are the same as ergot alkaloids. *A. fumigatus*, like all molds, thrives in wet conditions, but the optimum temperature for its growth is 37° C. This mold, rather than ergot, may have been the causal agent in epidemics of central nervous system disease in hot weather.[20]

Prior to 1650 it was commonly believed that a supernatural being, benign or malign, was the cause of ergotism. In the century that witnessed the beginning of the scientific revolution, however, many educated people began to seek natural causes for the symptoms. New labels were given to the disease—"hysteria," "vapours," "hypochondria," "nervous fever," "fits," and "frenzy." But the search for natural causes did not succeed for over a hundred years.

Confusion was understandable, for the symptoms of ergotism varied from one time to another, from one place to another, and from one case to another to a much greater extent than the symptoms of most diseases. Ergot was not one substance but a package of chemicals, the content of which varied according to differences in climate, topography, the nature of the plant host, and other factors. It was enough to baffle the most intelligent investigator.

Among the ergot alkaloids, however, are red dyes that give distinctive tints to white bread made from rye infected with ergot. In 1864 N. Zinin said that in Russia bread with an ergot content of 0.5–3.0 percent was pale pink; 3–5 percent, pink; and over 5 percent, red. In 1927 in Perm province bread having 2 percent ergot content was violet, that with 3–6 percent was cherry colored with a violet shade. Color provided a way of identifying ergot in refined rye flour products, but not in unbolted dark rye products.[21]

It was observed that ergot was most likely to form on rye when the preceding winter was cold. Severe weather generally traumatized plants. Another favorable condition was a cloudy and wet spring, which lengthened the time the rye flowers were open and hence vulnerable to infection (ergot spores "take root" in the rye flower). Maximum alkaloid production occurred when air temperature was between 17.4° and 18.9° C.[22]

A local volcanic eruption producing much dust and ash (tephra) may also stimulate ergot infection of rye. A dust pall blocks out light from growing food plants, the air cools, and fogs form, maintaining constant humidity. When the sky is overcast the rye flowers remain open longer, making them more vulnerable to infection. Moreover, *Claviceps purpurea* thrives in constantly damp conditions.

Also favorable for ergot infection of rye was the presence of ergot-infected grasses growing wild in nearby areas. This was more likely to be the case when the land was newly cultivated. Low, shady, marshy land provided more water for the growth of *Claviceps purpurea*. Soils that were not deeply plowed permitted the sclerotia to germinate and form spores in the spring. Winter rye was more likely to be infected than spring rye.

In a community that grew both rye and wheat, consumption of rye was likely to increase after a severe winter, for winter rye, planted in the fall, was more likely to survive than winter wheat. But after a severe winter, rye was likely to contain more ergot than usual.

Ergotism epidemics could occur at any time because ergot alkaloids are very stable and may retain their toxicity for up to eighteen months. They do not break down during baking at low heat or boiling up to three hours. Home brew made from ergoty rye in Russia retained the poisonous alkaloids, but distilled alcoholic drinks did not.[23]

Ergotism epidemics usually occur after the rye harvest, however, when ergot is most toxic. This pattern is borne out by a drop in the fertility rate. In the population in western Europe dependent on rye, a trough in conceptions appeared in August or September after the harvest. One study has shown that in a year of high wheat prices in England, fertility began to decline within three months after the harvest, indicating early fetal loss.[24] One explanation for this drop might be that those who could not afford wheat or good rye were eating cheap, ergoty rye.

FUSARIUM TOXINS

The second group of mycotoxins of interest here are trichothecenes, produced by various species of *Fusarium* molds that contaminate cereals. Some species of *Fusarium* can produce toxins lethal to human beings.[25] The best-described form of *Fusarium* poisoning is alimentary toxic aleikiia (ATA), so named by the Soviet Ministry of Health in 1943. In 1934 N. E. Murashkinskii showed that a species of *Fusarium* infecting cereals was responsible for the disease. It is now known that several *Fusarium* species produce deadly toxins in all major cereal crops.

Alimentary toxic aleikiia as observed in the Soviet Union was more lethal than ergotism. (See Table 2 for a list of symptoms.) The agent responsible is called "T-2 toxin." Rice is the best biosubstrate for molds producing T-2 toxin, but rye, wheat, white corn, barley, oats, yellow corn, and millet may also be infected (in that order of suitability as a substrate). In an episode of T-2 toxin poisoning in 1944 in the Autonomous Tatar Republic in the Soviet Union, death occurred mainly among those who had eaten poisonous rye products.[26] Potatoes overwintered in fields, unlike grain, did not become poisonous.[27]

People of all ages may suffer from ATA. One study found the illness was most likely to strike people between the ages of eight and fifty years, with few cases among young children.[28] Another report, however, put 42.3 percent of all deaths in 1944 in one region of Orenburg in east Russia among children between two and ten years of age. Nursing infants were normally unaffected because *Fusarium* toxins, unlike ergot toxins, can not pass through mother's milk.[29]

The discrepancy between these reports may be explained in two ways. First, it is unclear whether the authors were drawing upon hospital populations or on the population at large. Many Russian families were reluctant to hospitalize sick children of tender age, so hospital-based statistics would underrepresent this age group. Second, there may have been regional differences in method of preparing food for young children. Those children fed only on porridge were less likely to die than those fed bread, pancakes, or noodles. In any case, although reports varied on the age distribution of the sick, it was agreed that the case mortality rate was highest among children under five and the elderly.

People on a monotonous cereal diet, low in proteins, were especially susceptible to ATA. Typically two-thirds of all cases occurred in May, but in Orenburg in 1942 July was the peak (57.6 percent of all cases). Of those who suffered an acute attack of the disease in the Soviet Union in the 1940s, half died.

T-2 toxin both in lethal and sublethal doses may cause immune deficiency. The pluripotential stem cells of the bone marrow may be injured, affecting all subsequent cell populations. Not only cells that do battle with foreign invaders but also those responsible for coagulation are affected. This accounts for both necrosis and the tendency

TABLE 2. Symptoms of alimentary toxic aleikiia

EARLY SYMPTOMS

Mild cases
  headaches
  listlessness
  chill
  fleeting pains in back, joints, sides, etc.
  nausea
  vomiting
  copperish taste or pepperish smart in oral cavity

Severe cases
  diarrhea
  profuse sweating
  fluttering, weak pulse

ACUTE PHASE

Mild cases
  tonsils and uvula painfully inflamed and swollen; possible presence of white specks
  putrid smell about the patient
  stool dark and very fetid
  skin eruptions: red flush, miliary eruptions

Serious cases
  widespread swelling and necrosis in the oral cavity, with difficulty in swallowing and a putrid breath
  skin eruptions of dark red color with large, distinct, pale pustules and sometimes purpuric spots, which tended to spread with the least scratch or trauma
  bleeding from the nose, mouth, and vagina; bloody urine and stool
  possible swelling of the glands in the neck, armpits, and groin; swelling of the neck glands together with swelling in oral cavity might cause strangulation
  abnormal appearance of blood
  vomiting, diarrhea, profuse sweating
  fever
  pulmonary complications
  central nervous system disorders: delirium, stupor, convulsions, depression, disorientation
  meningeal hemorrhage

to bleed. The cause of the immune deficiency may be overlooked, however, because attention is naturally directed to the infections invading the body. Immune deficiency may also result when T-2 toxin attacks lymphoid cells. Such cells are critical for the defense of the body against infection.[30]

There are a number of reasons for believing that in the past ATA was often mistaken for an infectious and contagious disease. Like infectious diseases, ATA was familial in character and caused fever in its victims. But in other respects ATA does not resemble infectious diseases. Its victims may suffer relapses, especially if they are reexposed to the toxin. Nursing infants and those on a milk diet are not vulnerable, because a sick human mother or a cow does not secrete in her milk T-2 toxin, the principal causal agent in ATA. And unlike tick- and louse-borne relapsing fever, ATA is characterized by throat ulcerations and other necroses resulting from opportunistic infections.

Because of these symptoms of the throat, ATA was sometimes confused with diphtheria, or diphtheria mixed with scarlatina, but in cases of diphtheria a false membrane forms in the throat, there is no bleeding, and in temperate climates there is no skin eruption. Today scarlatina is thought to be characterized by fever, skin eruptions, and sore throat, but not by an ulcerated throat or bleeding.

ATA may also be confused with scurvy, but scurvy does not affect the throat. ATA is similar to pellagra in its timing (spring and early summer) and some of its symptoms—skin eruptions, diarrhea, and mental disturbances. But ulcerations of the throat and hemorrhagic symptoms are not associated with pellagra. In recent years sporadic cases of ATA may have been diagnosed as aplastic anemia or agranulocytosis of unknown etiology.

Milling, cooking, and storage have little or no effect on the toxicity of grain infected by *Fusarium tricinctum,* a producer of T-2 toxin. Low-temperature baking has little effect. The safest food made from infected grain contains no husks and is boiled more than thirty minutes to make porridge. Only 2.8 percent of the sick studied in the Tatar Republic had consumed toxin in the form of porridge; the others had consumed it in bread, pancakes, or noodles. The toxin remains in undistilled alcoholic beverages, such as beer brewed from infected grain, but not in rye whiskey. Consequently, epidemics may continue over many months until the grain and its products are consumed or discarded.[31]

The toxin starts forming in grain in the fall but reaches peak toxicity in spring. Its formation is favored by a winter with a temperature in the 10°–15° C range, with abundant snow and alternate freezing and thawing. It was thought that maximum toxin formation occurred at temperatures between 5° and 14° C, but recently it was reported that a *Fusarium* strain from South Africa produced a large amount at 25° C. Sharp fluctuations in temperature serve to increase toxicity. There is no relationship between the rate of growth of the fungi and the toxicity of the grain. Long after the fungi disappear, the toxin remains—up to seven years after the grain harvest.[32]

Grain that is late to mature and contains over 16 percent moisture is especially vulnerable to infection. So is overwintered grain, left out in the field uncut or scattered or in small heaps, and grain grown in black, loamy soils and clays and in low-lying areas (below 350–400 meters altitude).

T-2 toxin is not the only dangerous *Fusarium* toxin; there are others that damage the human immune system.[33] In recent experimental animal studies scientists have identified toxins that can suppress the action of neutrophils and B-lymphocytes, which fight bacteria, and T-helper lymphocytes, which protect the body from mycobacteria (causing leprosy and tuberculosis), fungi (causing mycoses), protozoa, and viruses (causing smallpox, measles, influenza, and more). Cats are particularly susceptible to oral doses, but adult poultry, swine, and rodents are less susceptible than are adult human beings.[34]

In addition to the products of genus *Fusarium*, other mycotoxins have immunosuppressant capabilities, such as aflatoxin, produced by the mold *Aspergillus flavus*. Molds that attack the immune system are found in corn, nuts, oil seeds, and other foods.[35] Thus, someone who eats contaminated food but does not die of acute poisoning may eventually be killed by an infection because of an impaired immune response. The immunosuppressive effects of certain mycotoxins may serve to explain some of the bizarre diseases recorded in history that fit no known disease model.

Mycotoxins may also be involved in the etiology of cancers, cardiovascular diseases, and birth defects.[36]

Recently scientists have discovered mycotoxin contamination of stored grain even in developed countries. In 1982 Basil Jarvis reported that in a study of breakfast cereals in the United Kingdom he had found that sixteen out of twenty-seven samples tested had a detectable level of trichothecene toxin, the greatest contamination being in wheat-based products.[37] In another study *Fusarium tricinctum* was found in cereal in Italy. Using corn as a substrate, the researchers produced T-2 toxin from the fungus in the laboratory.[38] Gabriella Sándor reported the presence of T-2 toxin in feed samples during 1977–1982 in Hungary; in the peak year (1980), 28 percent of samples contained this toxin.[39]

Contamination with T-2 toxin has been reported in Germany.[40] In the United States there is widespread occurrence of ergot, aflatoxin, and other mycotoxins in field crops.[41]

METHODOLOGY FOR HISTORICAL RESEARCH

The methodology for the study of fungi is now well established, but the study of past epidemics of mycotoxicoses is new. The difficulties in such research are great. Past observers reported symptoms selectively, in accordance with their preconceptions and limited by the state of knowledge of their day. Before the eighteenth century they tended to report only dramatic or bizarre symptoms. Systematic reporting of nationwide epidemics to the government began in France only in 1776, in England in 1837, and in most other places in the late nineteenth century. These statistics cannot always be taken at face value.

Equally difficult to study is the influence ergot alkaloids and other mycotoxins may have had in episodes of bizarre behavior. The first aim of investigation must be to get the most detailed and precise descriptions possible. The most valuable kind of account is one kept by a local physician and published as a monograph or diary. Diaries kept by the clergy are also helpful because many clergymen had medical training and attended the sick in their parish. Court records may be the only source of information about bizarre behavior in cases involving alleged bewitchment.

Since direct evidence of mycotoxicosis may be difficult or impossible to find, I have had to rely on indirect evidence to gauge the possible effects of poisoning on demographic trends and historical events.

The argument from incidence. Seasonal peaks in mortality and troughs in conceptions serve as circumstantial evidence for the presence of mycotoxins. This argument is used throughout the book.

The argument from prevalence. A second argument is based on regional differences in diet and climate that are associated with regional differences in fertility and mortality. This is the argument used in Chapter 2.

The argument from exposure. Class differences in diet (for example, if the rich eat only white wheat bread) or ethnic differences in diet (if the Irish eat only potatoes with little or no cereals) may explain class and ethnic differences in fertility and mortality. This argument is used in Chapter 9.

The argument from elimination. A dietary change may result in a marked change in fertility and mortality trends. The gradual elimination of rye, its replacement by wheat, maize, and potatoes, and the consequences are described in Chapters 7, 11, and 12.

When it *is* possible to quantify causal factors of the phenomena I discuss (as in Chapter 2 on Russian fertility and mortality), the numbers themselves specify the relative importance of each causal factor (such as temperature or wheat prices). To appreciate those numbers, a knowledge of elementary statistics is necessary. If it is impossible to quantify causal factors, which is usually the case, then merely mentioning "other possible causes," which are beyond the scope of this research, adds little to the discussion.

Current theories fall short in explaining changing fertility, changing mortality, and consequent population trends in the past. Neither famine nor infectious diseases are sufficient causes of mortality trends. There is an alternate explanation available, however: the operation of poisons produced by fungi. In this chapter I have shown how these poisons work and how they can have widespread effects. The next chapter will present a case study.

The reasoning I developed to explain demographic trends promises also to shed light on particular historical episodes that are only imperfectly understood. Along with the rise in population in Europe from 1750 to 1850, there was also a remarkable improvement in the mental health of ordinary people. Epidemics affecting the central nervous system, frequent before 1750, became rare after 1800. Hallucinations, convulsions, and panic still afflicted individuals, but rarely were whole communities affected simultaneously. Scapegoats were created to take the blame for outbreaks of abnormal behavior, but they were not hanged or burned for witchcraft. Instead of spreading the word that the Kingdom of Heaven was at hand and then waiting for a miracle, people joined in organizations to attain specific political goals.

In the chapters to come I will extend my argument that the presence of mycotoxins in the food supply of a population may have caused episodes of bizarre behavior. Indeed, it is my position that ergot poisoning was entirely responsible for the appearance of abnormal central nervous system symptoms in the communities examined in this book. (Of course, ergot poisoning had nothing to do with how those symptoms were interpreted—as bewitchment, say, or as divine inspiration.) The psychotic symptoms of fungal poisoning are well documented. Now we must ask whether these same symptoms match the description we have of symptoms in the past, before dietary change occurred.

CHAPTER TWO

A Case Study: Russia and Its Neighbors

PRIOR to 1945 the Russian people suffered more from fungal poisoning than any other major European nation. The root of the trouble was the cold Russian climate, which favored mycotoxin formation and forced the Russians to depend on rye as a dietary staple. Moreover, knowledge of the dangers of mycotoxins was limited, even among the educated. Doctors were unable to diagnose ATA until the 1930s: they referred to its symptoms as *gnilaia zhaba* but did not know the disease was caused by food poisoning.

The earliest diagnosed epidemic of ergotism in Russia occurred in 1785–1786 along the Desna River, between Kiev and Chernigov, around the town of Oster.[1] A physician writing in 1797 correctly identified "black horns" on rye as the cause of epidemic symptoms of the central nervous system. He associated the appearance of ergot with moisture and low elevation and with a wet, dark, cold summer in which the harvest was delayed. The author was familiar with the writing of the Swiss doctor Simon-André Tissot on ergotism (1765, 1778–1780). By 1841 the Russian government was aware that a parasitic fungus was responsible for the production of ergot.[2]

Whatever they were called, epidemics of ergotism were probably indigenous to Russia from earliest times. In the lives of Russian saints written during the fifteenth through seventeenth centuries, frequent references were made to epidemic nervous disorders among the masses. Widespread gangrene was also reported. These epidemics were associated with periods of famine.[3] In secular sources, occasional epidemics were recorded—in 1710 and 1722, for example—but reporting was very haphazard prior to 1832.

The first explicit warning about ergotism may be the one that appeared in 1798–1799 in *Derevenskoe Zerkalo*, a manual for "improving landowners."[4] Warned of such symptoms as a ravenous appetite and diarrhea, farmers were advised either to discard ergoty rye or to clean it by flotation. In 1828 the first detailed clinical description of the symptoms of ergotism appeared in Russia.[5]

The popular name for the disease was *zlaia korcha* (the evil writhing). Medical professionals might call in *rafaniia*. A single ergot, one of the "black horns" infecting the rye, was called *sporynia*, or *babii zub* (woman's tooth—which does not speak well for the dental health of peasant women). The term *nervous fever* might refer to ergotism, but not necessarily: other evidence beside this label must be present as well if we are to confirm the diagnosis. For example, in 1823 a report in a medical journal associated "nervous fever" with bad food and with reproductive disorders and bleeding from the vagina; this collection of data would support a diagnosis of ergotism.[6]

Ergotism was probably endemic in most parts of European Russia, but surveys of the disease pattern (called medical topography) were taken in only a minority of the areas affected. Viatka province (especially Sarapul district) was mentioned as early as 1835 as a focus of ergotism, called *porcha* by the common people and *hysteria* or *nervous fever* by physicians. In 1849 the government warned against feeding young children black or chewed-up bread, as was the practice in this province. It was estimated that the disease accounted for 25 percent of all deaths and that 20 percent of newborns died from it. Ergotism was blamed for endemic dystonias in children, often fatal.[7] Endemic dystonias were also reported in Kazan and Kiev provinces in the early nineteenth century and in Vologda in 1871.[8]

An indication of the demographic effects of the disease may be found in a report of an epidemic in the province of Perm in 1896. If a pregnant woman got ergotism when she was in her first trimester she aborted spontaneously; if in her sixth or seventh month of pregnancy, the child died after birth.[9] According to one authority, the average case mortality rate from ergotism was 20 percent but might run up to 70 percent.[10]

Government efforts to prevent ergot poisoning were limited to the army. In 1758 a systematic summary of earlier food hygiene measures, which remained authoritative for the remainder of the century, provided that the rye prepared for soldiers should be dry and be without any kind of strange admixture or smells, neither old nor musty.[11] After the widespread ergotism epidemics in 1832, the Ministry of Internal Affairs sent out to the various provinces directions based on government practice in Saxony for preventing the disease. In 1838, and again in 1863, scientific commissions were assembled to investigate the problem, but in 1882 it was concluded that there was no way to clean contaminated flour. Thereafter the central government took no action in the civilian sector—that was left to the *zemstvos* (the organs of local civilian government). In 1881 the permitted ergot content of rye for military use was set at 0.5 percent and in 1891 this level was reduced to 0.06 percent, where it remained into the Soviet period. (The permitted legal content in the United States is 0.03 percent).[12]

Only a fraction of the actual amount of ergotism that occurred was reported.[13] Years of particularly widespread reporting of ergotism were 1832, 1837, 1863–1864, 1909, 1926, 1928–1929. No epidemics were reported during 1905–1907, when revolutionary disturbances upset the normal functioning of government, yet in these years the weather was favorable for ergot formation. Again, during the period 1914–1924, years of war, revolution, and governmental reorganization, no ergotism epidemics were reported, although the weather was favorable for ergot formation from 1915 to 1919.

Given the deficiencies in reporting, it is likely that the significance of the disease in the history of Russian health has been seriously underestimated. The reasons for underreporting during war and revolution are obvious; but even in peacetime underreporting was typical because ergotism was a rural disease, affecting mainly the most unfortunate members of rural society. The number of scientifically trained observers in poor rural areas was insignificant prior to the founding of zemstvos in the late nineteenth century, and very limited from the late nineteenth century until the 1930s.

Improvement in the situation probably began, however, before World War I. Increased potato consumption reduced dependence on cereals, and potatoes were not prone to insidious infection.[14] They also served as a hedge against rye harvest failure, so that fewer peasants were forced to eat moldy and ergoty grain in time of famine. In the Baltic states and Belorussia potatoes became a staple human food after 1850, perhaps contributing to the relatively low mortality of this region.[15] By 1912 over half of the potato crop was used for human and animal consumption, the remainder being used for distilling and seed. Central Russia had become more dependent upon potatoes for food, but in southern Russia the potato was eaten only rarely.[16] In the early twentieth century (1896–1915), per capita consumption in kilograms of potatoes was 22.2 percent of the total diet.[17]

There are no data on potato consumption trends in Soviet Russia, but potato production increased from 31.9 metric tons in 1913 to 49.4 in 1930, and 76.1 in 1940, and it has since varied between 67 and 97 metric tons a year.[18] It is likely that a result of increased potato consumption since the early twentieth century has been a decline in mortality from ergotism. Nevertheless, ergotism may have made an important contribution to the death toll as late as the 1930s. Although no statistics were published, it is believed that mortality was high in the Ukraine from 1932 to 1934; perhaps man-made famine was not entirely responsible.[19] The toll was especially high in the year 1933, and in the following areas: the Ukraine, Northern Caucasus, Middle and Lower Volga, and Kazakhstan. Perhaps ergotism, the incidence of which usually rises in time of dearth, accounted for part of the death toll. The following chronology, composed on the basis of previously unpublished government data, was published in a Soviet botanical journal in 1939.[20]

1932. There was "significant" development of ergot on rye; 30–50 percent of ears were infected. The affected areas included Belorussia, Leningrad province, the northern part of Ivanovo province, the north and central regions of Moscow province, Gorki province, the northwestern part of the Tatar A.S.S.R., Povolzhia (now Saratov) province, Voronezh province, the northwestern part of the Ukraine, and parts of Ordzhonikidze region.

1933. A very bad year: mass infection of rye ears occurred because there was a large stock of ergots in the soil from the preceding year and because the wet weather of spring and summer in 1933 favored ergot formation on the new crop. On the average 75 percent of all rye ears were infected in Belorussia and the Ukraine and in Ivanovo, Yaroslavl, Moscow, Smolensk, Voronezh, Krasnodar, and Ordzhonikidze provinces. Infection averaged 50–75 percent in Gorki (formerly Nizhni Novgorod), Vologda, Archangel, and Leningrad provinces and on the Lower Volga. Infection averaged 20–50 percent in the Bashkir and Tatar republics, in the Urals, and in Odessa and Kuibyshev (formerly Samara) regions.

1934. This was a dry year, not favorable to ergot formation, in central and southern Russia, and only in the north and northwest was ergot formation above normal. Over 75 percent infection occurred in Leningrad, Smolensk, Moscow, Ivanovo, Yaroslavl, Voronezh, and Kiev provinces, in Belorussia and around Alma-Ata. The infection rate was 50–75 percent in Gorki, Archangel, Vologda, Saratov, Stalingrad, Kursk, Chernigov, Krasnodar, Kuibyshev, Cheliabinsk provinces and the Tatar A.S.S.R.

1935 and 1936. Ergot development was insignificant.

After 1936 the only bad ergot year was 1938. The areas that suffered most in this year were Vologda, Kalinin (formerly Tver), Kirov (formerly Viatka), Perm, Cheliabinsk provinces, the Bashkir A.S.S.R., and in Siberia, Novosibirsk, Omsk, the Altai, and Krasnoyarsk regions. During World War II the infection dropped to less than one-tenth that of 1938 because less rye was sown. After the war improved agricultural techniques kept the ergot content of rye harvests at a safe level.[21]

How important was the influence of mycotoxins on Russian health? Statistical evidence for the Soviet period is unavailable, but it is possible to study health in late Imperial Russia, 1861–1913. To be sure, there is little direct evidence of the causes of death in Imperial Russia. Surveys are available only for cities, and these only on an irregular basis.[22] We must therefore approach the problem indirectly, using statistics on the birth rate, death rate, crop distributions, and mean monthly temperature.

It is clear that before World War I the health of the Russian people was extremely poor. In 1897, the year of the first national Russian census, the infant mortality rate for European Russia was 260 per 1,000 births, compared to 222 for Germany, 164 for France, 156 for Italy, 156 for England and Wales, and 109 for Ireland.[23] The crude death rate in 48 provinces (*gubernias*) of Russia (excluding Moscow and St. Petersburg provinces) was 32.1 per thousand, a level comparable to the average for seventeenth-century England. Yet the population was increasing because of high crude birth rate: an astonishing 49.5 per thousand.[24] Why?

There is no reason to believe that the peasants were unusually fond of sexual activity but, confined to their huts during the long, severe winters, they may have been more active during that season than peasants in milder climates. Russian peasants married early—typically girls married at eighteen years, boys at nineteen, compared to twenty-five and twenty-eight, respectively, in northwestern Europe. The Russian teenage husband and wife were able to start having children immediately because they did not try to set up an independent household: they started married life in an extended family household, sharing responsibilities with the husband's parents, brothers, brothers' wives, and sisters.[25]

There is no basis, however, for inferring that Russian family structure was mainly responsible for the high fertility: such a structure may equally well be interpreted as an adaptation serving to offset the high infant mortality rate. In 1904 it was estimated that 10 percent of the difference between fertility rates in Russia and those in Western European countries could be explained by the different proportions of women between the ages of fifteen and fifty who were married, and 10 percent more could be attributed to differences in the age structures of populations.[26] That left 80 percent of the variance unexplained.

The cause of the high crude death rate and the high infant mortality rate in Russia is also obscure. There is no reason to suppose that unusually unsanitary conditions and undernourishment were responsible: the English poor were probably just as dirty and ill-nourished as the Russian poor in the nineteenth century. The poor of both countries depended heavily on a diet of starches and were quite often, and quite involuntarily, vegetarians.[27] English villages usually lacked public bathing facilities, but in the typical Russian village there was a sauna (*bania*).

It has long been assumed that English public health improved in the last quarter of the nineteenth century mainly because the English cleaned up their water supply. There is of course some truth in this: two severe water-borne diseases, cholera and typhoid fever, subsided at that time in England. In Russia typhoid continued to be a problem in cities, and sanitary measures to protect public health were inadequate. But this contrast turns out to be weak: in Russia 80 percent of the population was rural and dispersed, and water-borne diseases therefore must have played a less important role there than in urbanized England.

So other explanations for Russian health problems must be sought. I undertook the present inquiry with the knowledge that ergotism and alimentary toxic aleikiia were associated with certain climatic conditions, particularly cold winters. The logic of inquiry was that (A) variation in temperature and diet predicted the occurrence of (B) mold poisoning. These were likely to influence (C) fertility and mortality, especially infant mortality. The occurrence of (B) could not be measured, but if (A) predicted (C), and no other explanation was available, then (B) might be supposed to be the intermediate variable.

The dependent variables in the multiple regression analysis were the crude birth rate, the crude death rate, and the infant mortality rate as reported in the years 1896, 1897, and 1898, and the means of these variables for 1861–1913.[28] For the census year 1897 the fertility rate corrected for the proportion of women between the ages of fifteen and fifty who were married was used. The crude death rate of 1897 was corrected for age by multiplying the total population of each province by a coefficient representing its variation from the national mean in the proportion of persons up to nine years of age in the population of that province. The result was then divided by total deaths to give the age-corrected death rate. Because the two diseases studied were known to be common in rural, but not urban areas, the provinces of St. Petersburg and Moscow were omitted from the study, leaving 48 provinces.

Of the predictor variables used, mean January temperature was expected to correlate negatively with mortality, especially infant mortality, since cold winters predicted both ergotism and ATA. Mean April temperature was expected to correlate negatively with mortality, predicting ATA only. Mean July temperature was originally included as a predictor variable because in western Europe it was known to be positively correlated with deaths from infectious diarrhea, and the results of the regression might aid in differential diagnosis. It had little or no significance, however, in predicting mortality in Russia and high significance in predicting fertility. Mean May and June temperatures were also tested, and May temperature was found to be predictive of both fertility and mortality. June temperature predicted neither. May, June, July, and August levels of precipitation were tested and found to have some predictive value, until these variables were introduced into a multiple regression analysis, where they were "swamped" by the effects of the more powerful temperature predictors.

The reports of 238 weather stations were followed,[29] at least one being in each of the 48 provinces studied. Multiple observations in a province were averaged. In the case of missing data, an observation was used for that year from an adjacent province with a similar climate.

A high proportion of rye in the total starch production of the 48 provinces was expected to correlate positively with mortality from ergotism, whereas a high proportion of potatoes was expected to correlate negatively with both forms of food poisoning.[30] Some rye was grown in every province of Russia. Even in the south, where large crops of wheat were harvested, most of the wheat crop was exported and the peasants depended on rye. For the very reason that rye was ubiquitous, it was not a good predictor variable.

Phenological phenomena were considered in the choice of precipitation measurements: the wetter the weather in May and June, when the rye was in flower, the greater the chance of ergot infection; the wetter the weather in July and August, the higher the moisture content of the harvest and the more likely it would be infected with *Fusarium* strains when stored.[31]

Another predictor variable was the incidence of breast-feeding, which provided infants protection from *Fusarium* toxins but not from ergot poisoning if the mother was suffering from ergotism. Supplementary cereal feedings of nursing infants increased in July and August, when women were busy with agricultural labor. This was especially true if the men were away working in St. Petersburg or Moscow.[32] Summer was indeed the peak season for mortality in the period studied.[33]

Weaned infants and small children were particularly susceptible to mold poisoning, as they consumed a greater amount of food in proportion to body weight than did adults. In the ergotism epidemic of 1855–1856 in the German region of Hesse, for example, 50 percent of victims were under the age of ten; in the epidemic of 1862 in Finland the percentage was 56 percent.[34] Thus, when women who worked the fields in the summer weaned children early, they exposed them to the danger of disease.

The amount of breast-feeding was measured in two ways. One was an index of differential male/female migration to the capital cities; the assumption was that the greater the proportion of males who left, the more the women in the villages would be forced to perform summer field labor.[35] Another measurement was devised in recognition of the fact that the Bashkirs and Tatars, who constituted a majority of the population in Kazan and Ufa provinces, mandated breast-feeding of infants for a longer period than did the Russian people. They did not use contaminated cereal pacifiers. The proportion of these ethnic groups in the southeastern Russian provinces was thus in itself a way of predicting mortality.[36]

FINDINGS

A map of the distribution of mean crude death rates in the 48 provinces studied, 1861–1913, shows a concentration of high rates in the northeast (Figure 3). This is the coldest part of European Russia.[37] In the multiple regression for the crude death rate for this period the independent variables found to be significant were the January and May temperatures, the amount of potatoes grown as a percent of total starches, and the percent of Bashkirs and Tatars in the population (here labeled "ethnicity"). The relative weights these variables had in accounting for variance in the crude death rates (the dependent variable) in the different provinces are indicated by the multiple regression coefficient r^2 and the F-statistic.

Variable	r^2	F-statistic
January temperature	0.48	43.0
May temperature	0.64	41.1
Percent potatoes/all starches	0.67	29.5
Ethnicity (percent Bashkirs and Tatars/total population)	0.70	25.0

If the crude birth rate were added to the predictor variables, r^2 for the crude death rate would rise to 0.881, or 88.1 percent (F score of significance = 31.1). *This means that the factors listed above could account for 88.1 percent of the variance in mortality.* This is a highly significant result.

A map of the distribution of infant mortality rate (Figure 4) again shows a concentration of higher mortality in the northeast. Kazan and Ufa had anomalously low mortality rates, but otherwise there was marked discontinuity between areas of high mortality in the northeast and lower mortality in the west. The contributing variables to the regression were as follows.

Variable	r^2	F-statistic
January temperature	0.60	68.5
Ethnicity (percent Bashkirs and Tatars/total population)	0.65	41.3
Percent potatoes/all starches	0.68	31.2
Percent rye/all starches	0.73	28.4
Migration (male/female migrants to capital cities)	0.76	26.3

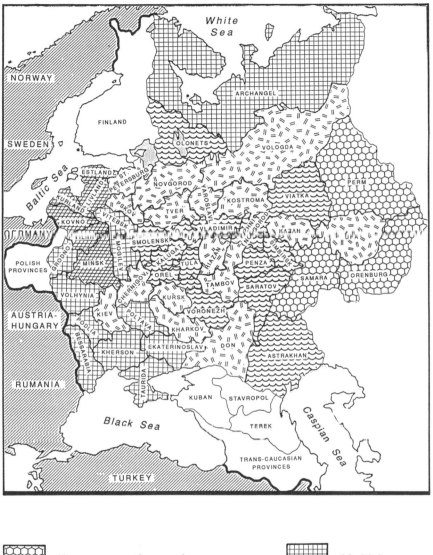

40 or more per thousand		26–30.9	
36–39.9		25.9 or less	
31–35.9			

FIGURE 3. Mean crude death rates in European Russia, 1861–1913.

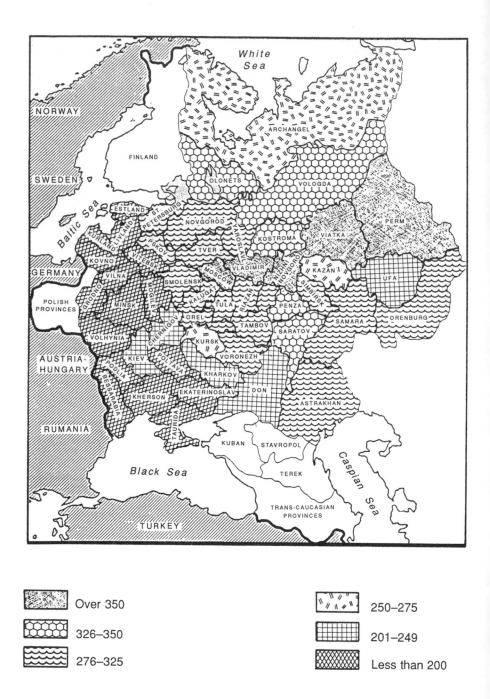

Over 350

326–350

276–325

250–275

201–249

Less than 200

FIGURE 4. Mean infant mortality rates in European Russia, 1867–1910.

If the crude birth rate were added as a predictor variable, r^2 for infant mortality would increase to 83.2 percent (F = 20.9). *This means that the factors listed above accounted for 83.2 percent of the variance in infant mortality.* This, too, is a highly significant result.

A scattergram of the strongest predictor variable, January temperature, and infant mortality (Figure 5) showed a concentration of high mortality rates in provinces with mean temperature in the −10° to −15° C range. The provinces fell into two groups, those with low infant mortality rates in the warmer west and south, and those with high infant mortality rates in the cold north and east. This pattern served to support the view that *Fusarium* toxins were important agents in infant mortality

A map of the distribution of the crude birth rate (Figure 6) shows the focus of high fertility to be the southeast, where summer temperatures were highest. The low fertility of the Baltic states was associated with cool summers, moderate winters, and low mortality. The multiple regression for the crude birth rate yielded these significant predictor variables:

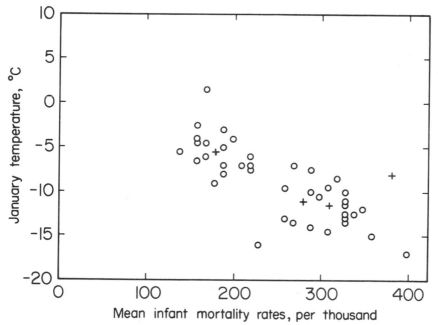

FIGURE 5. Infant mortality and January temperatures, 1861–1913; a cross indicates a collision mark for more than one score of the same value.

Variable	r^2	F-statistic
July temperature	0.30	19.2
January temperature	0.57	29.4
Migration (male/female migrants to capital cities)	0.61	23.3
Ethnicity (percent Bashkirs and Tatars/total population)	0.64	18.7

If the crude death rate were added to the analysis, r^2 would be 85.5 percent ($F = 24.9$). *This means that the factors listed above explained 85.5 percent of changes in fertility.* Again, this is a highly significant result.

There was an unexpected negative correlation between January temperature and fertility ($r = -0.424$), which was also observed in the studies of 1896, 1897, and 1898. Scattergrams showed that the relationship between these variables was linear for values greater than $-7°$ C, suggesting that in unusually cold winters increased sexual activity may have been responsible for 18 percent of the variance in fertility.

The year 1896, immediately preceding the census of January 28, 1897, was one of average mortality (31.8 per thousand) but colder than average mean January temperature, $-12.2°$ C (the average was $-9.0°$ C), and April temperature, 2.1° C (the average was 4.8° C). The regression predicting the age-corrected death rate showed the overwhelming influence of the two variables ($r^2 = 0.649$). Mean January temperature alone predicted 49.5 percent of the variance in infant mortality.

In 1896 mean July temperature, which was slightly above average, was negatively instead of positively correlated with the crude birth rate ($r = -0.56$), whereas May temperature, which was also slightly above average, had a slightly weaker positive correlation with the crude birth rate (0.41 instead of 0.48). January temperature had a strong negative correlation (-0.75) with the crude birth rate.

The year 1897 was one of below-average mortality and average winter cold. January temperature alone predicted 53.3 percent of the variance in the age-corrected death rate and 50.3 percent of variance in infant mortality. Fertility was measured by the "overall fertility" index calculated by Coale and colleagues[40] on the basis of the 1897 census, which took into account the proportion of women who were

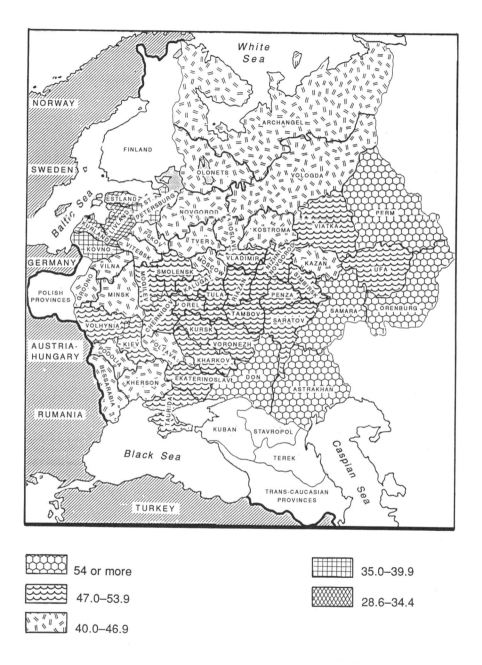

White
Sea

NORWAY

FINLAND

SWEDEN

ARCHANGEL

OLONETS   VOLOGDA

Baltic Sea

ESTLAND

ST. PETERSBURG

PERM

NOVGOROD   VIATKA

KOSTROMA

LIVLAND

PSKOV   YAROSLAVL

KURLAND

TVER   NIZHNI NOVGOROD

KOVNO   VITEBSK

VLADIMIR   KAZAN

UFA

GERMANY

VILNA   SMOLENSK   MOSCOW   SIMBIRSK

GRODNO   MOGILEV   KALUGA   TULA   PENZA   SAMARA   ORENBURG

POLISH
PROVINCES   MINSK

OREL   TAMBOV   SARATOV

CHERNIGOV   KURSK

VOLHYNIA   VORONEZH

AUSTRIA-
HUNGARY

KIEV   POLTAVA   KHARKOV

PODOLIA

KHERSON   EKATERINOSLAV   DON   ASTRAKHAN

BESSARABIA

RUMANIA

TAURIDA   KUBAN   STAVROPOL

Black Sea   TEREK

Caspian
Sea

TURKEY   TRANS-CAUCASIAN
PROVINCES

54 or more   35.0–39.9

47.0–53.9   28.6–34.4

40.0–46.9

FIGURE 6. Mean crude birth rates in European Russia, 1861–1913.

married and between the ages of fifteen and fifty. This corrected measure of fertility correlated with May 1897 temperature (0.68), with July 1897 temperature (0.58), and with May 1896 temperature (0.67). The multiple regression containing these variables, plus January temperature in 1896 and 1897 and the crude death rate of 1896, predicted 84.4 percent of the variance in fertility.

Using the results of the regressions with the 1861–1913 population means and the 1881–1915 temperature means, I constructed models for predicting the three dependent variables: crude birth rate, crude death rate, and infant mortality rate. I then studied data for the year 1898 and compared the observed values with those predicted by the models. The crude death rates in 1898 were corrected for age by multiplying by this fraction: (number of persons aged 0–9 years in the province, 1898)/(population of province in 1897 + 18y), where y = average monthly growth of population, 1897–1914.

When all predictors for 1898 were used, the r^2 for variance in the age-corrected death rate was 87.1 percent. January temperature alone predicted 66.1 percent. The r^2 for variance in infant mortality was 86.2 percent. January temperature alone predicted 72.6 percent. The r^2 for variance in the crude birth rate was 77.8 percent; January temperature and May temperature combined predicted 62.1 percent of the variance among provinces.

The expected mean age-corrected death rate was 33.7 per thousand population; the observed mean was 32.2. The expected infant mortality rate was 236 per thousand births; the observed rate was 268. The expected crude birth rate was 42.8 per thousand population; the observed rate was 47.2.

The climatic data for the year 1909 were incomplete, but partial data suggested that it was an average year. The proportions of potatoes and rye in the total starch yield were determined (Table 3), and it was learned that between 1885 and 1909 there was a steady increase in the proportion of the potato crop. In the year 1909 the amount of potatoes grown had no significant effect on infant mortality (1908–1909 mean), but it was negatively correlated with crude death rate (1906–1910 mean) (r = −0.53). *In the early twentieth century increased potato consumption could account for 28.1 percent of the reduction of mortality rates.*

Several explanations for the population changes in Russia described here must be considered. One is that infectious diseases were the primary cause of death, but the evidence of the seasonality of

TABLE 3. Means of variables in Russia, 1885–1909

Year	Crude death rate (per thousand population)	Infant mortality (per thousand population)	Crude birth rate (per thousand population)	Potatoes as a percent of total starches	Rye as a percent of total starches
1885	34.1	274	46.7	17.5	40.8
1896[a]	31.8	259	48.1	27.5	30.7
1897	30.7	259	48.7	31.4	28.7
1909[b]	28.6	238	45.8	35.3	31.0
1913	27.1	n.d.	43.9	29.2	20.9
1926	19.1	172	43.5	37.4	20.6

Sources: Russia, Tsentral'nyi Statistichcskii Komitet, *Dvizhenie Naseloniia v Evropeiskoi Rossii za 1885 g.* (1896 g., 1897 g.) (St. Petersburg); Russia, Tsentral'noe Statisticheskoe Upravlenie. *Statisticheskoe Obozrenie,* 1927, no. 1 (January), pp. 28–29; B. R. Mitchell, *European historical statistics, 1750–1975,* 2d ed. (New York, 1980), pp. 129 and 141; A. G. Rashin, *Naselenie Rossii za 100 Let* (Moscow, 1956), pp. 167–68, 187–88, 195–96.

a. 1896: Mean of 1892–1896 data for crops.

b. 1909: Mean of 1906–1910 data for crude birth rate and crude death rate; mean of 1908–1910 data for infant mortality.

death rates does not support this hypothesis. Peak mortality in Russia occurred in the summer. The widespread acute infections of the time, except for tuberculosis, cholera, infectious diarrhea, and typhoid, were winter diseases. Tuberculosis of the lung peaked in the spring; it was common among army recruits and young adults in St. Petersburg, but it could not account for the high rate of infant mortality. Of the summer diseases, epidemic cholera had subsided in Russia in the 1870s. After a more localized outbreak of 1892, it went into abeyance until 1926. Moreover, although the incidence of cholera tended to increase with greater summer temperature, in Russia summer death rates bore little relationship to temperature.[39] Typhoid fever probably contributed to adult mortality, but it was not a disease of infants.

Diarrhea may be caused either by infectious pathogens, which are more active in hot weather, or by food poisoning, which is associated with the seasonal crop cycle. It is unlikely that such pathogens were wholly or even mainly responsible for summer infantile diarrhea in Russia, because the number of deaths was not correlated with summer temperature. In London during the period 1840–1909 infectious diarrhea peaked in the summer and had a positive correlation coefficient of 0.62 with mean summer temperature. In Russia from 1861 to 1913, however, July temperature did not contribute to the multiple regression analysis for infant mortality rate and was correlated only slightly ($r = 0.28$) with the crude death rate.

Moreover, in Russia the temperatures of the preceding January and April had a powerful influence on mortality rates in the summer months. This can be explained if the effect of January and April temperatures was not to hasten the spread of infectious diseases that caused deaths the next summer but to increase the likelihood that food supplies, grown or stored in those months and ingested the following summer, would be poisoned. This logic appears to be confirmed by a recent report that intestinal disease in Russia, peaking in the summer, was in large part noninfectious in nature.[40]

As for other possible causes, famine was not a problem in Russia except in the year 1891, and then only in the dry southeastern part. Death from starvation probably contributed little to mortality rates. Nor was Russia engaged in a major war during the years studied.

This leaves the possibility that food poisoning was a major contributor, if not the most important contributor, to the great summer surge in deaths, especially the deaths of infants and young children. The findings presented here generally support such an inference. Mean January temperature, a predictor of both ergotism and ATA, was the most important predictor variable in most mortality regressions. Less important, but also significant, were April and May temperatures, differential male/female migration rates, and the share of potato production in total starch production. July temperature, important in determining ergot alkaloid production, was a strong predictor of natural fertility.

The unexpected positive correlation between the crude birth rate and mean May and July temperatures prompted further investigation. As corroboration of this finding, the crude birth rate of Finland over the period 1751–1850[41] was correlated with an index of spring and summer temperature: the width of tree rings in Lappland (northern Finland) for the same period.[42] The correlation coefficient was 0.56; if the war years 1808–1809 were omitted, $r = 0.60$. This should be compared with the 0.41 correlation for May and 0.55 for July in European Russia of 1861–1915. (In the period 1751–1850 Finland was heavily dependent on rye as staple food, and epidemics of ergotism were frequently reported.)[43]

Two factors may be needed to explain fertility variance in ryedependent lands: difference in *Claviceps* strains and difference in climate. In 1983 a team of researchers in Kiev[44] tested 529 strains of *Claviceps purpurea*, of which 44 percent produced ergot alkaloids. Each strain produced a characteristic spectrum of alkaloids, the average amount of ergonovine being 5–10 percent of the total alkaloids. This was relatively low in relation to the amounts produced by *Claviceps* strains in other European countries.

Climatic conditions mainly influenced the total amount of alkaloid production. A team of Slovenian investigators reported in 1981 that the lowest yields of ergot alkaloids under field conditions occurred when temperatures were 19.5°–23.6° C. Maximum yields were obtained when temperatures were 17.4–18.9° C.[45] The number of rainy days, rather than the amount of precipitation, was important.

Using a scattergram for July temperature (means, 1881–1915) and the crude birth rate (means, 1861–1913), I found that the data points clustered around the line of regression when temperature was between 16.0° and 20.8° C. In a scattergram for May temperature and the crude birth rate of 1897, data points clustered around the regression line when temperatures were between 16.0° and 19.4° C.

Researchers working in the Soviet Union in the 1950s and 1960s found the highest alkaloid output in the *Claviceps* strains growing in the forest and forest-steppe zone of Russia (Figure 7).[46] In this zone of high-alkaloid-producing *Claviceps* strains, only the central, non-black-soil region (such as around Orël, Kursk, Voronezh, Tambov, Penza, Tula, and Kaluga provinces) had the cold, wet climate favorable for ergot development. In the cold northern provinces (Archangel, Vologda, Yaroslavl) and in the wet western region (the Baltic States and Belorussia), climatic conditions were most favorable for ergot development but the *Claviceps* strains there had low alkaloid yields.[47]

In summary, variance in temperature seems to influence the severity of *Claviceps* infection of rye, the total number of ergots produced, and hence the total production of alkaloids. Variance in *Claviceps* strain seems to determine the proportions of various alkaloids in each ergot. The central nonblack-soil region of Russia, at the core of the Muscovite state, was cursed with both the "right" climate and the "right" *Claviceps* strains for high alkaloid yields. This would explain why the Muscovite Russian population did not begin to increase rapidly until it spread out beyond this region.

There has been no general mapping of various *Fusarium sporotrichiella* species in European Russia. In 1971 a study was made of the distribution of *Fusarium* strains, 744 in all, in the Altai, Stav-

CONTENT OF
ALKALOID IN % AGE

0–0.005
0.005–0.05
0.05–0.1
0.1–0.2
0.2–0.3
MORE THAN 0.3

FIGURE 7. Alakaloid yields of *Claviceps* strains in Russian provinces.

ropol, North Kazakhstan, and Orël regions;[48] 83.7 percent were in the T-2 toxin-producing *F. sporotrichiella* group. There was no significant difference in the distribution of strains in the different areas tested.

Were Finland and Russia exceptional among rye-dependent countries in their "sensitivity" to climatic variables? Did the relationships observed in Finland and Russia also prevail in central and western Europe when those areas were rye-dependent and without crop surveillance systems? A comparative study was undertaken of twelve rye-dependent European countries, using data from the earliest available twenty-year period in the nineteenth century (Table 4). The results suggest that the patterns described in this chapter were once widespread in Europe north of the Alps and Pyrenees.

The evidence of Table 5 suggests that variance in January temperature was once an even more powerful predictor of fertility in western and central Europe than it was in Russia. In addition, it should be mentioned that two other countries, Italy and Ireland—which grew little or no rye—had exceptionally low crude death rates in 1897: 21.9 per thousand for Italy and 18.5 for Ireland.

In western European rye-dependent countries prior to moderniza-

TABLE 4. Means of variables in Europe, 1751–1880

Country	Years	Crude birth rate (per thousand population)	Crude death rate (per thousand population)	Mean January temperature (° C)	Mean July temperature (° C)	Rye as a percent of total starches
United Kingdom	1751–70	34.0	27.3	2.8	15.9	n.d.
Norway	1751–70	34.4	25.9	2.8	15.9	16.9
Sweden	1751–70	34.9	27.4	−4.1	17.4	29.1
Finland	1751–70	43.8	30.2	−6.1	17.4	n.d.
Denmark	1820–39	30.7	22.5	−1.4	17.0	18.8
France	1801–20	31.9	26.9	3.1	19.0	14.9
Belgium	1830–49	32.2	25.7	2.2	17.5	10.4
Netherlands	1840–59	33.5	25.9	2.0	19.0	17.5
Austria	1820–39	38.8	30.4	−2.2	20.7	19.2
Germany	1840–59	35.5	26.6	−2.7	18.8	21.0
Hungary	1861–80	42.4	37.0	−1.1	22.2	15.8
Russia	1861–80	50.3	36.6	−9.0	19.3	47.0

Sources: B. R. Mitchell, *European historical statistics* (New York, 1980), pp. 90–113 and 209–263; E. A. Wrigley and R. S. Schofield, *The population history of England, 1541–1871* (Cambridge, Mass., 1981), pp. 531–535.

TABLE 5. Correlations between variables in Europe and Russia

Variables	Correlation coefficient	
	$N = 12$ countries	$N = 48$ Russian provinces
Crude birth rate/ Crude death rate	0.895	0.786
January temperature/ Crude death rate	−0.558	−0.695
January temperature/ Crude birth rate	−0.793	−0.424
July temperature/ Crude death rate	0.440	0.543
Percent rye/ Crude birth rate	0.773	0.169
Percent rye/ Crude death rate	0.525	0.473

Sources: see Table 3.

tion endemic infection of rye by ergots that produced less lethal but fertility-suppressing alkaloids may have served to maintain lower levels of both mortality and fertility. Perhaps when the natural fertility of Europe is investigated, an appropriate question would be, "Why was the natural fertility of western Europe so low?" rather than "Why was the natural fertility of Russia so high?"

A similar reorientation might be proper in comparing the low fertility of seventeenth- and eighteenth-century England with the high fertility of colonial America. In America maize served as a staple starch and rye consumption tended to decline in the early eighteenth century. South of New England hot summers minimized alkaloid production of *Claviceps*. Instead of asking why the natural fertility of the American colonists was so high, perhaps one should ask why that of England was so low.

The amended hypothesis resulting from this study is that, given the distribution of microfungal strains and climatic variance of rural European Russia, a low mean January temperature predicted abundant ergot formation on rye ears, and a low April temperature predicted T-2 toxin formation in overwintered and stored grain. These poisons tended to increase mortality, especially among infants and young children, during the following summer. A mean temperature range of 17.4° to 18.9° C in May and July, which was optimal for ergot alkaloid formation, tended to be correlated with lower fertility. Future research may confirm the hypothesis that in Russia, and perhaps in other rye-dependent parts of Europe, epidemics of fungal poisoning served as the main limit on population growth before 1900.

In western Europe before the nineteenth century, health conditions were just as poor as those in Imperial Russia. In part 2 I explore the health history of various regions, beginning with the period of the Black Death, 1348–1350, in Europe.

PART II

Contributions to a Health History of Europe

CHAPTER THREE

A New Look in the Distant Mirror

The staff of life was perversely the scepter of death during plague
epidemics.
—MICHAEL DOLS, 1977

1390. A great mortality increased in Norfolk and in many other countries
in England, and it seemed not unlike the season of the great pestilence: it
was occasioned by a great want of fictuals that forced many people to eat
unwholesome food, and so brought distempers upon them.
—FRANCIS BLOMEFIELD,
An essay towards a topographical history of the Country of Norfolk,
vol. 3, 1805–1810

THE HIGH MIDDLE AGES,
the name usually given to the twelfth and thirteenth centuries, were
generally healthy, a time of rapid population growth. This, at least,
is the consensus of scholarly opinion. No statistics are available for
this period, to be sure, and medieval annalists who mentioned diseases did not usually describe them in enough detail to permit an ex
post facto diagnosis. All that is known is that the frequency of reporting of epidemics was much lower than that of later centuries.

The Black Death of 1348–1350 was an unprecedented demographic disaster. Instead of recovering in the next generation, Europe
suffered from a continuing demographic depression that lasted until
about 1490. In addition, the Late Middle Ages, as the fourteenth and
fifteenth centuries are called, were punctuated by epidemics of bizarre behavior.

THE PANDEMIC OF 1348–1350

Of all the epidemics in Europe prior to the sixteenth century, that of 1348–1350 was the one most fully described by contemporaries. There can be little doubt that it caused a severe demographic setback: about one-third of the population was lost. When a death toll is that large it seems appropriate to consider that widespread damage to the human immune system may have been involved. In other words, we must ask if the death rate was so high because of the virulence of the infectious agent alone, or if the ability of the human body to fend off disease was impaired in some way.

There is no experimental evidence linking the occurrence of plague with immunosuppression in humans, but it is well known that the case mortality rate from plague may vary widely in neighboring communities. In twentieth-century Russia, for example, it was 96.2 percent in Astrakhan province (1909–1910) and only 26.7 percent in Odessa (1910).[1] We must consider whether immunosuppressant mycotoxins in grain could account for these differences.

Graham Twigg has suggested that some epidemics labeled "plague" could not have been bubonic plague. Even in 1348–1350, supposedly the time of the worst bubonic plague epidemic, some observers reported necrotic and hemorrhagic symptoms that would indicate not plague but an immune-deficient condition.[2] Furthermore, a poisoned food supply could have increased the incidence of plague in other ways besides weakening humans' resistance to infection. Mycotoxins in the *rats'* diet may have been just as damaging, for an increase in rat deaths may increase the number of plague-infected fleas tapping human beings in place of their former hosts.

A brief review of the symptoms of plague seems in order. Its nonspecific symptoms include shivering, fever, vomiting, headache, giddiness, intolerance to light, pain in the abdomen, back, and limbs, insomnia, apathy, and delirium. Its most characteristic symptom is a bubo—a hard, painful, hemorrhagic swelling of a lymph gland. The buboes are usually found in the groin, but sometimes in the armpit or the neck. A bubo, ranging in size from an almond to an orange, is palpable the first day, and conspicuous the second to fifth day, of the disease. There is a marked hemorrhagic tendency—petechiae, the spots that indicate hemorrhaging under the skin, are common— which is currently presumed to be the effect of plague toxin.

A flea that infects human beings, *Pulex irritans*, can serve as the carrier of the disease, but the most common vector is another flea,

Xenopsylla cheopis, which infests rats specifically. The pathogen of plague, the bacillus *Yersinia pestis*, can infect fleas. It begins the epidemic cycle when it enters the stomach of a flea that has bitten an infected animal and dined on its blood. In the stomach of the flea the bacillus multiplies rapidly until the stomach is filled with bacilli. At this point the flea is called a "blocked flea." If the rat host dies of the bubonic plague, or anything else, the flea will have to find another host.

When the flea bites a human and sucks its blood, it regurgitates blood and plague bacilli into the bite site. Blocked rat fleas can stay alive at least six weeks. The optimal temperature for fleas is 20° to 25° C, and if the ambient temperature rises above 27° C the block is dissolved and the flea is cleared. Flea larvae cannot tolerate relative humidity below 60 percent. Heat and drought will shorten the life of a blocked flea; cold, dry weather is believed to cause it to hibernate. The plague bacillus itself is not sensitive to cold and can survive for twenty-five years if it is stored below 20° C.[3]

An examination of historical records shows some well-established associations between occurrence of plague and the following variables: age of victims, the presence of stored grain, and environmental moisture. The reasons for these associations are not entirely clear.

Age. Plague occurs most frequently in young people. In a study of the ages of plague victims in London in 1603, Mary and T. H. Hollingsworth found that plague mortality differed from normal mortality mainly in higher death rate of children (5–14 years) and youths (15–24 years).[4] (This study may be criticized on the ground that in normal times the deaths of children were so greatly underreported that child mortality from plague only appeared excessive.) Leslie Bradley, in a study of the epidemic of 1665–66 at Eyam, England, found that mortality from plague among infants (0–1 years) was much below average; among children, 1–4 years, below average; and among those aged five to twenty-four years, mortality was three times normal.[5] In 1424 in Tuscany, infants were fairly immune to plague, but 69.1 percent of all who died were children.[6] In a study of 38 cases of plague in the American West, 1970–1980, J. M. Mann found that 55 percent of these cases were children sixteen years old or younger.[7]

The age of plague victims is relevant to the argument because the young, *depending upon how fast they are growing, how active they are, what they are eating, and how the food is prepared,* may ingest more mycotoxins per unit of body weight than their elders do. Contradictory results relating to the age incidence of plague may eventually be explained by taking such factors into account. Mortality may be higher among young men than among children, for example, if the children are eating porridge boiled long enough to break down mycotoxins (over thirty minutes), or if young men are in the midst of harvesting or other heavy labor and therefore require more calories per unit body weight than usual.

Grain storage. Plague tended to appear where there were surpluses of grain—and, because grain is the rat's favorite food, a "surplus" of rats. Along with their hosts, rat fleas also thrive in cereal debris, a suitable pabulum for their larvae. Stored grain surpluses tended to be found in commercial areas, such as southeastern England and along the Rhine and the Rhône rivers. The medieval population was heavily dependent on cereals.[8] All small grains, including wheat (the staple of the Levant) and rye (the staple of European commoners north of the Alps and Pyrenees and east into Russia), can serve as substrates for mycotoxin formation.[9]

Environmental moisture. The occurrence of plague is very strongly correlated with the incidence of humidity, rain, and flooding. In 1975 J. N. Biraben concluded:

> We have found innumerable examples where rain has provoked a recrudescence of the plague, beginning with the year 1348, which is mentioned by the chroniclers as very rainy and humid in Mediterranean regions. . . . We have never found a contrary example, where the plague ceased following rains. . . . All the causes of dryness stop the plague, and we can cite many examples.[10]

For two years prior to the pandemic of 1348 in Europe the weather was extraordinarily rainy and humid and the crops poor. The summer of 1348 was exceptionally wet in England, where the plague began its ravage; on the other hand, it was not so wet in Scotland that year, and the plague did not spread widely there until the wet summer of 1350.[11]

The coldest and driest regions were untouched by plague during the pandemic. Iceland, northern Norway and Sweden, Finland, and large areas of Russia and the Balkans escaped the blight. So did the mountainous and desert areas of the Near East. Plague probably spread to eastern Europe via rivers and inland seas, which were natural routes of commerce, but it did not fan out into surrounding territories that were cold and dry.

Ralph Josselin, an Essex clergyman, kept a detailed diary giving information on the weather and on plague fatalities during 1665–1666. He reported that the grain harvested in 1664 was very wet. The following fall was rainy. During February and March there was alternate thawing and freezing. (These conditions were ideal for the production of T-2 toxin in stored grain.) On May 14 he reported the outbreak of plague in London. For the remainder of 1665 and into 1666 plague mortality was closely tied to rainfall.[12]

One wet country that suffered relatively little from plague in the sixteenth and seventeenth centuries was Ireland. It may be significant that the Irish subsisted mainly on dairy products and depended little on cereals.[13]

This association between plague and moisture is puzzling in that *Rattus rattus* and its flea, *Xenopsylla cheopis*, both dislike and avoid dampness.[14] Indeed, water-soaked grain is unhealthy for rats. Rats prefer soft, finely divided food, especially the softer parts of grain. If grain is infected by fungi, rats avoid the infected parts. *But if grain is soaked in water, rats will eat it whole.*[15] If they do that they are more liable to ingest *Fusarium* toxins, and if all available grain is soaked in water, then the rats are even more liable to ingest toxins.

A possible explanation. Little is understood about the distribution of plague epidemics in space and time. Why were some areas exempt from plague while neighboring areas were afflicted? In the great pandemic of 1348–1350 some parts of Europe escaped harm altogether and others suffered relatively low mortality. The latter included Belgium, Alsace, Lorraine, Franconia, Bohemia, and parts of eastern Europe.

Graham Twigg has cast doubt on the widely accepted belief that bubonic plague was really pandemic in 1348–1350. He thinks that it was present only in Mediterranean ports and a few cities, where human and rat populations were dense, but that in most parts of Europe density was too low for the spread of the disease. As evidence he points out that in England, where cool weather conditions were unfavorable for the development of plague, the case mortality rate in the pandemic was relatively high.[16] Other aggravating factors, or other diseases, were probably necessary to cause such high mortality. If this premise is correct, it seems justified to shift attention from the invading pathogens of the plague to the defending immune systems of humans and rats.

It is well known that the microfungi producing immunosuppressant toxins. flourish in the moist conditions associated with plague epidemics. T-2 toxin in coincidence with plague epidemics may have markedly increased rat mortality from a variety of pathogens. The rapid spread and high mortality of plague during these years may have reflected injury to the natural protective mechanisms of the human *and* rat populations. If more rats died—for whatever reason—the likelihood of a rat flea feeding on a human being would increase. Indeed, the rat flea would be more likely to attach itself to *any* warm body. During the wet period 1345–1350 mortality among horses, cattle, sheep, and goats was exceptionally high.[17]

The food-poisoning hypothesis serves to explain not only the connection between moisture and plague mortality, but also the mortality rate by age. Plague tended to kill children and youths far more than mature adults. Individuals who are growing fast and expending much energy ingest more food per unit of body weight than do full-grown adults; if the food is poisoned, they ingest more poison per unit of body weight. Wholly breast-fed infants ingest no mycotoxins; newly weaned children are likely to be fed soft foods, like porridges, which have been boiled more than thirty minutes and thereby made harmless. In some parts of Russia children of tender age rarely suffered from alimentary toxic aleikiia (caused by T-2 toxin), and this may be because they were given boiled porridge, not bread.

The food-poisoning hypothesis also serves to explain higher plague mortality among the poor. Especially in times of dearth, the poor were forced to eat cereal products of questionable quality. To be sure, the rich could more easily escape infectious disease by moving away from stricken areas; but they also had access to better-quality cereals and were therefore less likely to have been infected with immunosuppressant mycotoxins.

The familial character of plague mortality in medieval Europe can be explained by the family's common source of staple food: contaminated grain. Differences in plague mortality in adjacent areas can be explained by differences in rainfall, diet, and storage practices.

In summary, increased levels of T-2 toxin and related toxins in grain, being harmful to the immune status of human beings and rats, may have contributed to the occurrence of the bubonic plague. However, it is also possible that other natural immunosuppressive agents, not of fungal origin, contributed to the high mortality of the pandemic of 1348–1350.

THE DEMOGRAPHIC DEPRESSION, 1350–1490

Oh miserable and very sad life . . .
We suffer from warfare, death and famine;
Cold and heat, day and night, sap our strength.
Fleas, scabmites and so much other vermin
Make war upon us.
In short, have mercy, Lord, upon our wicked
persons, whose life is very short.[18]

The French poet Jean Meschinot wrote these words more than a century after the Black Death had subsided, but the population of Europe had not yet recovered from the disaster. The depth of the demographic depression varied in Europe north and south of the Alps. In Tuscany (the only well-studied part of Italy) population was in a trough in 1430, but began to recover slowly after 1450. In northwestern Europe the deepest depression was probably between the 1430s and 1480s with a trough in the 1450s.

Table 6 contains some very rough approximations of demographic statistics calculated on the basis of population estimates of McEvedy and Jones and the chronology of "plagues" (not necessarily the bubonic plague) of Biraben. What it suggests—and this is to be taken with very great caution—is that conditions were much worse in Italy than in northwestern Europe during 1350–1430. After 1430 conditions in Italy improved, but not in northwestern Europe.

The number of abandoned villages in England shows that the peak in depopulation occurred between 1433 and 1486.[19] Walter Janssen found that the fourteenth century was disastrous in the Rhineland

TABLE 6. Public health in northwestern Europe and Italy, 1351–1499

	Estimated population, 1400 (millions)	Number of epidemics, 1351–1430	Epidemics per million population, 1351–1430	Estimated population, 1500 (millions)	Number of epidemics, 1431–1499	Epidemics per million population, 1431–1499
NW Europe[a]	26.75	625	23.4	36.95	710	19.2
Italy	7.0	235	33.6	10.0	220	22.0

Sources: Colin McEvedy and Richard Jones, *Atlas of world population history* (Harmondsworth, 1978); and Jean-Noël Biraben, *Les Hommes et la peste en France et dans les pays européens et méditerranéens* (Harmondsworth, 1978).

a. Northwestern Europe includes the British Isles, France, the Low Countries, Germany, Austria, Bohemia, and Switzerland.

in terms of the abandonment of farms and villages, the worst fifty-year period being 1450–1500.[20]

A recent study by Guy Bois shows that conditions were also bad in eastern Normandy.[21] Population levels hit three troughs, the deepest of which occurred between the 1430s and 1480s (Figure 8). Arlette Higounet-Nadal's study of Périgueux, in east-central France, also shows a population trough at this time (Figure 9).[22] In Brabant, Belgium, population declined from 92,738 in 1437 to 75,343 in 1496. In

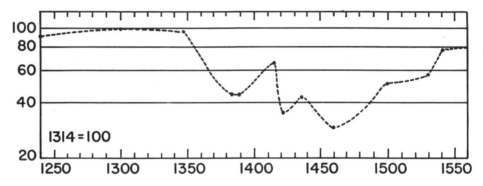

FIGURE 8. Index of population in eastern Normany, 1250–1550. Reprinted from Guy Bois, *The Crisis of Feudalism* (Cambridge: Cambridge University Press, © 1984), p. 76, by permission of Cambridge University Press.

FIGURE 9. Index of population in Périgueux, 1330–1490. Reprinted from Arlette Higounet-Nadal, *Histoire du Périgord* (Toulouse: Privat Toulouse, 1983), by permission of the publisher.

Basel, Switzerland, there was no population growth between 1429 and 1497, the trough falling in the 1450s.[23]

Two possible explanations for the demographic depression in northwestern Europe are famine and infectious disease. There were widespread famines in Europe during 1437–1439 and the early 1480s, but they were unusual. Given the one-third to one-half reduction in population that had occurred during 1348–1350, available supplies of land and domestic animals were usually sufficient. There was probably no lack of protein: in eastern Normandy even peasants enjoyed an abundance of salted fish, dairy products, and eggs the years round.

No doubt infectious diseases were widespread, but there are reasons to doubt that the pathogens of infection were chiefly responsible for the demographic depression. The denser a population, the easier it is for infections to spread. The density of population in most parts of Europe had been drastically reduced by the bubonic plague: mortality from infections should, by all rights, have declined. Second, R. S. Gottfried's study of East Anglia during 1430–1480 has shown that mortality varied greatly in neighboring regions and even communities. This pattern was not typical of the major infectious diseases.[24] Third, Europe was relatively isolated from the rest of the world: only a few explorers were exposed to exotic pathogens. Before the end of the fifteenth century, limited contacts abroad probably resulted in a negligible increase in the stock of pathogens in Europe.

Epidemics involving food poisoning, on the other hand, are tied to climatic and dietary conditions, and tend to vary as these conditions vary. From literary evidence there are some indications of changing levels of rye consumption during the fifteenth century. In 1438, a year of cold and famine, a London chronicler said that on account of the shortage of grain men who had never eaten rye bread and barley before were forced to eat it then.[25]

The elites of western Europe avoided rye bread, so one would expect them to fare better during ergotism epidemics. Ergotism epidemics can be distinguished by their sudden peak in late summer and fall. Gottfried found that during 1430–1480 mortality rates peaked in late summer and fall in East Anglia, but that the peaks of the mortality rates of the elites of both London and rural areas were less pronounced.[26]

Marie-Thérèse Lorcin found that among men affluent enough to make wills in Lyon there was a tendency to leave to the widow greater stocks of wheat than of rye. The proportion of wheat in total bread grains willed to heirs was as follows.[27]

Years	Percent of wheat
1300–1350	30.8
1350–1400	51.5
1400–1450	66.0
1450–1500	78.8

Perhaps the affluent of Lyon were growing increasingly aware of the danger of rye consumption as the climate turned wetter and cooler. Indeed, Lorcin found that the well-to-do left an increasing number of heirs: *they* suffered no sudden catastrophic mortality crisis in the fifteenth century.

Late medieval chroniclers left anecdotal evidence of epidemics of central nervous system disease among the commonfolk in northwestern Europe.[28] These diseases could have been caused by a number of toxins, among them ergot alkaloids. After a severe winter in the summer of 1340 in England, especially in Leicestershire, there appeared a lethal disease characterized by "fits," pain, and the compulsion to bark like a dog. In the summer of 1355 there was an epidemic of "madness" in England, both in town and country. Believing that they saw "demons," people ran into woods and thickets to hide or back and forth between their homes and fields.

Sometime between 1361 and 1365 (probably after the severe winter of 1362, 1363, or 1364) the St. Albans chronicler said that "Numbers died of the disease of lethargy, prophesying troubles to many; many women also died by the flux [probably vaginal hemorrhage]; and there was a general murrain of cattle." In 1374, a wet year marked by a dearth of food, there was an outbreak of hallucinations, convulsions, and compulsive dancing in the Rhineland. Some people imagined they were drowning in a stream of blood, which obliged them to leap high. This disease seemed to select children and the poor as its victims. In 1389 after a severe winter at Cambridge an epidemic of "phrensy" (with symptoms of the central nervous system and fever) arose suddenly and was very lethal.

In 1418 a similar epidemic came to the Rhineland. After a severe winter in 1458 a widespread dancing mania struck Germany. Several hundred children from Holland and Swabia set out on a pilgrimage for Mont St. Michel, and many died. The years 1480–1482, attended by famine and floods, were very sickly in the Rhineland and northwest Germany. After a severe winter in the summer of 1482 many people dashed their brains against a wall, rushed into the river, or ran to and fro before dying in agony. It may be that some or all of these episodes, which some attributed to abnormal psychology, had an organic basis, namely, food poisoning.

There is also indirect statistical evidence that ergot alkaloids, by reducing fertility, played an important role in the demographic depression of the 1430s–1480s. Indeed, health statistics for western Europe in later centuries suggest that ergot influenced population trends chiefly through fertility suppression. *Claviceps* strains in this region, unlike those of Russia, may have produced more fertility-suppressing alkaloids than lethal alkaloids.

The logic of the statistical argument may be presented as follows (the bracketed elements being unmeasurable).

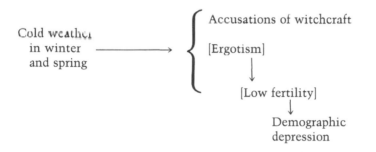

Cold weather in winter and spring ⟶ {
Accusations of witchcraft

[Ergotism]
↓
[Low fertility]
↓
Demographic depression
}

Indirect evidence of ergotism. A person suffering from ergotism may be prone to hallucinations, spasms, and twitches, and in the villages of Europe these symptoms were attributed to witchcraft. Although there is no index of the incidence of ergotism epidemics, there is an index of court cases of witchcraft accusation in Europe during 1300–1500.[29] This can be used as a proxy variable for ergot poisoning.

Evidence of climatic factors. Two useful climatic indices are available. An index of spring and summer temperatures is provided by Siren's measurements of Finnish tree rings. Lamb provides a crude index of winter temperatures from studies of English annals.[30]

The justification for relating late-medieval temperatures and fertility was given in Chapter 2. In rye-dependent states of nineteenth-century Europe, where ergotism was endemic, there was a positive correlation between temperature and fertility. Likewise in rye-dependent areas of late-medieval Europe, where records of witch trials provided evidence of endemic ergotism, but where fertility rates were unknown, one may consider temperature variance as a proxy for fertility variance.

What kind of climatic conditions prevailed during the demographic depression? In all but three years between 1436 and 1487

Finnish tree rings were narrower than average, signifying cold growing seasons. The coldest year was 1457. In the same time span English winters were colder than average in every decade except the 1470s. If the above reasoning is correct, then temperature and the rate of witchcraft accusation should be inversely correlated, for colder weather caused more ergotism and hence more symptoms of "bewitchment." And indeed such a statistical relationship exists. When the indices of summer and winter temperatures are regressed with the late-medieval index of witchcraft accusations, at lag 0 (temperatures of the same year) and lag 1 (temperatures of the preceding year), they predict 50 percent of the variance in accusations ($N = 150$, $F = 13.4$, very significant).

Late-medieval Europe also suffered from unusually wet weather. Such weather generally encourages the growth of fungi, including those that produce immunosuppressant mycotoxins. So one would expect to find a positive correlation between wetness and the occurrence of epidemics. The best predictor of epidemics in the British Isles during the period 1350–1489 (given five-year units) was summer wetness ($r = 0.634$, $N = 28$). (The data on both epidemics and climate were drawn from chronicles and collected in my personal file of natural catastrophes.) In the case of France from 1350 to 1479 (given ten-year units), wetness in summer and fall (as reported in Italy, data for France being deficient) was the best predictor of the so-called plague epidemics listed by Biraben ($r = 0.671$, $N = 13$). Cold winters ($r = 0.517$) and summers ($r = 0.567$) were also significant predictors. This evidence suggests that a causal connection may have existed between the cold, wet weather and the demographic depression of the late Middle Ages.

The late Middle Ages were a crazy and violent period, as Barbara Tuchman has shown. Perhaps the underlying causes for the tenor of the times were invisible. Perhaps medieval manuscripts do not tell the whole story. In this cold, wet period, various microfungi, competing with humans for food plants, may have had the upper hand.

CHAPTER FOUR

Mycotoxins and Health in Early Modern Europe

Europe climbed out of the demographic depression in the late fifteenth and early sixteenth centuries. In most areas in the seventeenth century growth slowed down or stopped. Population increased in the first half of the eighteenth century, but did not take off until about 1750. What prevented an earlier takeoff?

In the Early Modern Era population growth in the rye-dependent countries of Europe showed evidence of the restraining influence of ergot alkaloids. The best of this evidence relates to England, where ergot alkaloids apparently had little effect on mortality trends, but may have had a great influence on fertility. Other mycotoxins, causing damage to the immune system, may have contributed to mortality. This can be said because in the Early Modern Era doctors began to describe the symptoms of diseases in detail. They were able to find associations between the occurrence of certain diseases and environmental conditions.

Of particular interest were diseases characterized by necrosis, bleeding, and an ulcerous swollen throat, symptoms of damage to pluripotential cells in the bone marrow (today sometimes labeled "aplastic anemia"). Such symptoms indicate that the patient suffers from immune deficiency. While the causes of the condition cannot be known with assurance, its epidemiology suggests that mycotoxins may have been responsible, for outbreaks of immune-deficiency symptoms occurred in association with *food shortages* and unusually *wet* environments.

In the summer of 1661, for example, Dr. Thomas Willis described a bizarre epidemic in London. First the victims suffered from weakness, giddiness, loss of appetite, nausea, and weight loss lasting over many weeks. Then more severe symptoms appeared: bloody urine, coughing and lung problems, swelling of the lymphatic glands in the neck, flea-bite spots and ulcers on the skin, diarrhea, the formation of a whitish crust in the throat and mouth cavity, and central nervous system disorders. Whole families might be affected; children and the elderly were most likely to die. It may have been significant that there was a food shortage in London that year.[1]

Another seventeenth-century London physician, Thomas Sydenham, stated that in general, especially between spring and summer, the young were likely to suffer from "quinsy," a lethal suffocating throat disease, together with fever and the spitting up of blood.[2] Two English authors described the epidemiology of the bizarre disease characterized by necrotic and hemorrhagic symptoms. Thomas Short, after making a survey of the health history of western Europe, concluded in 1749 that the disease appeared after wet weather and was associated with famine conditions. The wet weather points to fungi; the famine conditions, to food poisoning. James Johnstone of Worcester reported in 1774 that "quinsy" was associated with warm moist weather and was most likely to appear after a mild winter. It occurred most often among poor people living in low, wet marshy locations, especially in a year of food dearth. Johnstone said that persons aged ten to fifty were most often sick, but that children were most severely affected.[3]

North of the Alps and Pyrenees such symptoms were reported only occasionally in the seventeenth century but more frequently in the first two-thirds of the eighteenth century, with a peak in the warm, wet 1730s. John Huxham, a physician in Plymouth, described the symptoms in detail, and William Hillary in Yorkshire, James Johnstone in Worcester, and John Fothergill in London provided confirming accounts.[4]

South of the Alps and Pyrenees the same symptoms were first described in detail in the sixteenth century. They were called "petechial fever," which was very lethal for children. Victims bled from the nose and had red or purple spots (petechiae) on arms, back, and chest.[5] Better described was the epidemic of 1513, which resulted in high mortality rates in England and Italy. This epidemic, too, was associated with rainy weather, as well as the consumption of unwholesome food. The victims were pale and had ulcers and black spots on their skin, diarrhea, difficulty breathing, black stools and urine, swollen feet, and a putrid smell. When physicians bled them the effect was negative.[6]

The same symptoms reappeared often in Italy and Spain in the first half of the seventeenth century. If the throat symptoms were present the disease was called *garrotillo.* It was widespread in the rainy year of 1629, but after 1651, although it continued to appear in southern Europe, it was less prevalent.[7]

The reoccurrence of the same types of symptoms, over several countries and several decades, must be more than coincidental. Medical reports were only sketchy, and the names for the various outbreaks of disease were hardly consistent, but the continual reference to necrotic or ulcerous symptoms, hemorrhaging, and the like indicate to me a common cause. I believe that cause was fungal poisoning of the food supply, which, as we have seen, waxes and wanes with weather conditions and grain availability. The same can be said, in my opinion, in regard to fluctuations in fertility.

ERGOT AND FERTILITY IN ENGLAND

Historical demographers have scrutinized England's rich records more than those of any other country. Of all western European countries, England provides the best evidence for the influence of mycotoxins on vital rates in the Early Modern Era. It seems likely that ergot alkaloids had an overwhelming influence on fertility trends there.

England at this time depended heavily on rye bread: this fact has been buried and needs to be unearthed. At the beginning of the seventeenth century small amounts of rye were grown in most parts of England, but large acreages were found in the northern counties (Cumberland, Westmoreland, Durham, Northumbria, Yorkshire) and in Shropshire, Cheshire, Herefordshire, the Petworth district of Sussex, Norfolk, East Suffolk, northwest and southeast Essex, and Dartmoor in Devon (Figure 10).

In 1584 Thomas Cogan[8] wrote that rye was not as good as wheat to eat because it was hard to digest; nevertheless it was "much used in bread almost throughout the realm." In the early seventeenth century Fynes Moryson[9] reported that

> The English Husbandmen eate Barley and Rye browne bread, and preferre it to white bread as abiding longer in the stomack, and

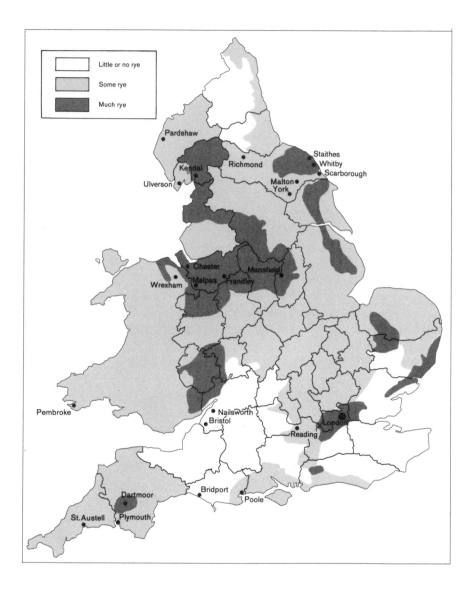

FIGURE 10. Distribution of rye cultivation in England and Wales. Sources: Eric Kerridge, *The agricultural revolution* (London, 1967); Joan Thirsk, ed., *The agrarian history of England and Wales*, vol. 4, *1550–1640* (Cambridge, 1967), p. 167.

not so soone digested with their labour, but Citizens and Gentle-
men eate most pure white bread, England yielding (as I have said)
all kinds of Corne in plenty.

After that time, however, dietary changes began to occur. By the
1620s the more prosperous farmers in the southeast (East Anglia,
Essex, and Sussex) were eating wheat or maslin products (maslin
being a mixture of wheat and rye), while at the same time producing
rye for the London market. The poor, who had difficulty paying for
rye bread in time of dearth, were resorting more and more to barley
and other substitutes.[10]

By this time the literate British citizens probably knew that wheat
bread was healthier to eat than maslin and that maslin was healthier
than rye. In 1620 Dr. Venner of Bath[11] wrote:

Bread made of Rie is in wholesomnes much inferiour to that
which is made of wheat: it is cold, heavy, and hard to digest, and
by reason of the massiveness thereof very burdensome to the
stomacke. It breedeth a clammie, tough, and melancholicke
juyce; it is most meete for rusticke labourers, for such by reason
of their great travaile, have commonly very strong stomacks. Rie
in diverse places is mixed with wheat and a kinde of breade made
of them, called Messeling [maslin] bread, which is wholesomer
than that which is made of Rie, but it is lesse obstructive, nour-
isheth better, and lesse filleth the bodie with excrements.

An anonymous guide to health published in 1624 stated that rye
flour, if well sifted, not made of whole meal, and newly baked with
meat, was good to eat in small quantities during the summer to
prevent constipation, and for that end was used at the tables of
princes.[12] By the middle of the seventeenth century a clergyman,
Ralph Josselin, while traveling in Leicestershire, was distressed at
the prospect of eating black bread and pocketed two white loaves
"against the worst." He continued to produce rye and maslin as well
as wheat at his home in Essex, however, and he and his family suffered almost every autumn (up to 1674) from "ague." Perhaps they
were mixing some rye flour with their wheat flour: for even those
who could afford pure wheat bread would do so, believing maslin
kept better than pure wheat bread.[13]

Beginning in the late sixteenth century, the English had imported
rye from the Baltic to meet the needs of the cities, especially London.[14] During the 1630s and 1640s, when these imports reached their

peak, fertility remained high, but around 1650 Baltic rye imports fell to one-twelfth their former level, and the difference had to be made up from English sources.[15]

From 1650 to 1750 England relied almost entirely upon its own cereal production. At first, from 1650 to 1682, wheat prices were high, and most people in central and northern England had to depend upon rye as a bread grain. After 1682, as agricultural techniques improved and wheat production increased, wheat prices began to fall. Gradually production of wheat, barley, and oats outpaced rye production, the process lasting throughout the eighteenth century.[16]

Ergot infection may once have been common in the rye crops of Britain. In 1980 G. Wood and J. R. Coley-Smith reported that ergot infection of grasses was currently prevalent throughout Britain and that most cereal varieties used in testing for ergot infection were regularly infected.[17] If ergot infection of grasses was prevalent in 1980, why should it not have been prevalent in rye in the seventeenth century, when that infection-prone crop was regularly and extensively cultivated? There is no evidence of any mutation in the varieties of cultivated rye (Secale cerealis) that would have increased its receptivity to ergot infection after the seventeenth century. Ergot infection of rye is global in scope; England, a trading nation since the medieval period, was not an isolated country in the seventeenth century, its crops protected from fungal invasion.

If ergotism was present in England, one wonders why English physicians did not suspect its presence. In 1765 the first essay on the cause of ergotism was published in English,[18] but it was not until 1785 that Johann Taube, a German, conclusively proved that ergot was the cause of the nervous form of the disease.[19] By this time rye consumption in England was rapidly declining. Since ergotism was a disease that tended to strike the poor hardest, and was best known as a disease of "backward" central and eastern Europe, the English were not eager to claim it as their own.

Charles Creighton, England's great medical historian, failed to diagnose most of the past ergotism epidemics. True, Creighton believed that three fourteenth-century epidemics and two early eighteenth-century epidemics with bizarre symptoms were ergotism. But he failed to recognize the disease in more commonplace seventeenth-century guises.[20]

Was there any clinical evidence of ergot poisoning in seventeenth-century England? Quite a lot. The earliest detailed account by an English physician of what may have been ergotism was that of Dr.

Edward Jorden in 1603. His brief treatise grew out of his involvement, the preceding year, in a witchcraft case in London, in which he testified that the defendant was innocent of the charge of bewitching the victim. Jorden declared that the alleged symptoms were those of a common disease called "suffocation of the mother."[21]

Jorden's disease "model" included intermittent disturbances of the central nervous system: tremors, spasms, chorea, convulsions, fainting, hallucinations, mood swings, compulsive laughing, catalepsy, paralysis; disturbances of the sense organs: temporary loss of sight, taste, feeling; speechlessness and a feeling of being choked or suffocated; disturbances of the gastrointestinal tract: vomiting and diarrhea; diverse pain; disturbances in body temperature and of appetite (either ravenous hunger or loathing of food); and "obstructions in the veins," perhaps a reference to vasoconstriction, ischemia (local and temporary blood deficiency), and formication (the feeling that ants are crawling under one's skin). The fit of these symptoms to the ergotism model is very tight. This conclusion is reinforced by reading the description of "suffocation of the mother" by John Sadler, a physician of Norwich, published in 1636.

By the middle of the seventeenth century English physicians were making an association between diet and central nervous system disorders. In 1653 Robert Pemell said such symptoms could be caused by bad breast milk.[22] (Ergot alkaloids can pass through breast milk to an infant.) In 1693 Walter Harris said that a woman of "hysterick" constitution should not nurse a child.[23] Brown bread was declared in 1695 to be bad for infants.[24] Thomas Sydenham said that an adult victim of "hysteric fits and colicks" should go on a diet of milk only.[25]

Even in normal conditions the natural fertility rate of England was low when compared with that of the continent during 1600–1799. There is no evidence of family limitation in England during these two centuries. Mean live births in "marriages of completed fertility" (a demographer's term to refer to marriages in which the wife has passed childbearing age) was 7.23. By contrast, in northwestern France (1670–1769) the level was 8.53; in northeastern France, 9.36; and in three Bavarian villages (1648–1849), 10.02.[26] Rye bread was a staple in both France and Bavaria, as well as England, but the *Claviceps purpurea* strains in England, compared with the fungal strains on the continent, may have been especially productive of fertility-suppressing ergot alkaloids.

English fertility varied with English rye and wheat prices. The higher the wheat prices, the more women ate rye bread and the less the fertility. The higher the rye prices, the more women had to eat ergoty rye bread and the less the fertility. A long series on rye prices is not available for England, but during the short period 1692–1703, rye prices were correlated ($r = -0.726$) with the crude birth rate in the year following ($t = -3.4$). In the case of wheat prices, the correlation was about the same ($r = -0.703$, $t = -3.1$).[27]

Another way to measure the influence of ergot on English fertility trends is indirect, through the use of climatic predictor variables. When these are used one assumes the action of ergot alkaloids to be an unmeasured intermediate variable.

Usable instrumental reading of temperature in England began in 1659.[28] The summer temperature readings up to 1723 were not accurate to 0.1° C, and because summer temperature varied within a narrow range, the readings of the era are of limited value. To estimate winter temperatures for the years 1645–1658 one may use data on the length of time the waterways in the Netherlands were frozen.[29] For summer temperatures before 1659 one must rely on estimates in annals[30] and on measurements of the width of tree rings in Scotland.

During the period 1647–1699, a time of low fertility, the mean winter temperature was 1.5° C, as compared with 2.1° C for the period 1700–1799. This sets off the second half of the seventeenth century as a time of colder winters. Summers tended to be warm enough, however, for ergot alkaloid production. Weather annals during 1645–1658 reported five hot summers (1645, 1651, 1652, 1653, 1654) out of fourteen. During the period 1659–1680 fourteen out of twenty-one summers had the minimum July temperature, 16° C, necessary for alkaloid production. But during 1681–1699, a time of increasing fertility, only three out of nineteen summers were warm enough for alkaloid production. If one selects from the period 1645–1699 those years in which cold winters were followed by summers warm enough for alkaloid production, one finds a total of nine, and all nine fell within the period of English fertility depression: 1641–1685.

English vital records were fairly reliable after the Restoration (1660)[31] and instrumental records of climate began then. A logical period for more detailed statistical study is that from 1660 to 1739, the last year before the mortality crisis of 1740–1742 and the beginning of the population explosion.

The economic variables (real wages, wheat prices, bread prices)[32] and temperature variables that predict the crude birth rate for 1660–1739 are shown in Table 7. Real wages were correlated ($r = -0.706$)

TABLE 7. Correlations of crude birth rate and crude death rate with economic and temperature variables in England, 1660–1739

		Crude birth rate		Crude death rate	
Variable	Lag	r	t	r	t
Real Wages	0	0.836	13.0	0.278	2.5
	1	0.829	12.7	0.263	2.3
Wheat prices	0	−0.777	−10.5	−0.438	−4.2
	1	−0.774	−10.5	−0.361	−3.3
Bread prices	0	−0.574	−6.0	−0.251	−2.2
	1	−0.568	−6.0	−0.260	−2.3
Winter	0	0.581	6.1	0.049	insignificant
temperatures	1	0.527	5.3	0.018	insignificant
Spring	0	0.448	4.3	0.396	3.7
temperatures	1	0.446	4.3	0.406	3.8
Summer	0	0.274	2.4	0.195	insignificant
temperatures	1	0.238	2.1	0.219	insignificant
Fall	0	0.378	3.5	0.361	3.3
temperatures	1	0.386	3.6	0.379	3.5

with wheat prices. These two were the strongest predictors of the crude birth rate. Winter and spring weather were next in importance, warm conditions favoring births.

The annual data were smoothed to eleven-year running averages (N = 120). All economic and temperature predictor variables were lagged one year. When seven predictor variables (real wages lag 0, winter temperature lag 1, spring temperature lag 1, wheat prices lag 0, crude death rate, bread prices lag 1, summer temperature lag 0) were used, r^2 = 0.924, F = 116. *This means that these variables could predict 92.4 percent of the variance in the English birth rate from 1660 to 1739, a highly significant result.*

This finding may be interpreted as evidence that ergot alkaloids had a strong influence on English fertility trends. In general, cereal dearth constrained choice in the quality of grain consumed, and cold weather, especially in winter, predicted ergot formation in rye.

MOLDS AND MORTALITY IN ENGLAND

One way to investigate causes of death is to focus attention on periods of peak mortality. In certain English mortality crises mycotoxins may have played an important role.

The year 1658 was one such crisis. The mean winter temperature (estimated at −1.0° C) was low for England and the spring unusually wet. In late August a great epidemic of twitching, dizziness, nausea, vomiting, diarrhea, fever, insomnia, spasms, convulsions, frenzy, cold and hot fits, headache, deafness, stupor, and skin eruptions was observed, often relapsing, especially in rural areas.[33] Oliver Cromwell suffered from insomnia, pain in the bowels and back, and "fits."[34] He died on September 3, 1658. A peak in mortality occurred in September. In the following year the birth rate fell. The period 1658–1659 was therefore characterized by the classic signs of an outbreak of ergotism.

The mortality crisis of 1741–1742 followed severe winter weather. Ulcerous throat symptoms were common in 1741; typhus appeared to have been common in 1741–1742.[35] Ergot poisoning does not appear to have played an important role, since fertility was only slightly below trend, but the action of immunosuppressant *Fusarium* toxins may be suspected.

Another way to investigate the influence of mold poisoning on mortality is to use temperature as a proxy predictor variable, as before. A statistical analysis of the English crude death rate, 1660–1739, reveals no single powerful predictor variable among those studied (Table 7). There is no evidence here that ergot alkaloids influence mortality trends, but the modest positive correlations of the temperature variables with the crude death rate are in accord with James Johnstone's observation that warm weather was associated with bleeding and necrotic symptoms. Such symptoms of immunosupression may have been evident in the case of acute disease, implying that subclinical immune deficiency was more widespread.

It is more difficult to explain the negative correlation of wheat prices with the crude death rate. Why did more people die when wheat was cheap? Was there something harmful about eating white wheat bread?

At any rate, from six predictor variables (wheat price lag 0, fall temperature lag 1, winter temperature lag 0, bread prices lags 0 and 1, and winter temperature lag 1) one may obtain a significant multiple regression coefficient for crude death rate ($r = 0.702$, $F = 27$). *This means that these variables predicted 70.2 percent of the variance in the crude death rate, 1660–1739.*

In summary, then, ergot alkaloids in rye bread probably had a strong if not a dominant influence on English fertility trends, while other molds caused ill health and early death for many. Yet the influence of molds on mortality was not so great in England as in Russia. This requires explanation.

England enjoys a maritime climate and does not suffer from temperature extremes. *Claviceps* strains there apparently did not contain highly lethal ergot alkaloids. Given proper methods it is possible to grow wheat in England, reducing dependence on rye. Moreover, in the Early Modern Era the English became a great commercial nation, able to vary their diet with imported food.

In the sixteenth century the English, both rich and poor, had about the same life expectancy at birth (35 years) as the Russians in the late nineteenth century. By about 1800, English life expectancy was increasing while Russian life expectancy probably was not. The changes in England will be the subject of Chapter 7. Meanwhile there is much to explore in the history of mental health in the Early Modern Era. Chapters 5 and 6 will be devoted to witchcraft persecution and other panics, which were manifestations, I believe, of ergot poisoning.

CHAPTER FIVE

Witch Persecution in Early Modern Europe

I<small>T</small> is difficult to think clearly about bizarre behavior, for it tends to arouse anxiety. Anyone who has ridden on a city bus with one or more harmless but obviously disoriented or disturbed persons will realize that others on board feel uncomfortable. Naturally people get even more uncomfortable when confronted with a whole crowd of psychotics. Even when one contemplates *past* insanity it is difficult to maintain objectivity. The urge to ride a hobbyhorse of some kind—moral, religious, intellectual, or political—is strong. To avoid this inclination one must define clearly what it is one is trying to understand. When witchcraft is the topic of investigation, the kinds of behavior that incriminated the accused must be identified and examined.

Anyone who has consulted the records of witchcraft persecution realizes that actual harm was done, and the harm fit a pattern. "Outbreaks" of witchcraft were often accompanied by outbreaks of central nervous system symptoms: tremors, anesthesias, paresthesias (sensations of pricking, biting, ants crawling on the skin), distortions of the face and eyes gone awry, paralysis, spasms, convulsive seizures, permanent contraction of a muscle, hallucinations, manias, panics, depressions. There were also a significant number of gangrene cases and complaints of reproductive dysfunction, especially agalactia (inability of a nursing mother to produce enough milk). Animals behaved wildly and made strange noises; cows too had agalactia. Not every victim of "bewitchment" had all the symptoms, but most had abnormal experiences and behaved in abnormal ways.[1]

The victim of persecution was a person accused of causing these symptoms in another. These symptoms were real. Certainly it was possible to "frame" someone for witchcraft, but that does not mean that all episodes of bewitchment can be ascribed to invention. Many of those affected were young children who could not be accused of feigning possession or having malicious plans to "get even" with a neighbor, and many died of their symptoms. The cause of harm was not known, but harm there was.

This pattern of symptoms, furthermore, was distributed in a nonrandom way in space and time. The characteristics of the nonrandom distribution will be described below. Only by linking the pattern of symptoms with a distinct epidemiology can we hope to find a reliable explanation.

The investigators who have looked into witchcraft persecution have proposed no adequate explanation for its epidemiology. They have not attempted to explain why witches were persecuted in one place and not another, at one time and not another. It simply will not suffice to discuss widespread beliefs about witches, tensions between factions in a witch-persecuting community, ruling-class repression, or legal and judicial arrangements for dealing with witchcraft. These were continuous cultural and social realities that did not vary in space and time as the distribution of "bewitchment" varied. One cannot explain a variable with a constant.

Of course, beliefs about witches, laws, court procedures, and social structures have been studied and should be studied to understand the social response to cases of "bewitchment." How the symptoms were perceived and interpreted was *culturally mandated.* They were seen through a screen of preconceptions.

Moreover, the persons accused of witchcraft were not chosen at random; nor were their persecutors from a random selection of community members. The existing body of witchcraft persecution theory tries to explain these nonrandom distributions of witches and persecutors of witches, but it does not explain the nonrandom distribution of witch-hunts. To blame witch accusers and courts for witchcraft persecution is to mistake an effect for a cause. Witches were persecuted because harm had befallen a community, not just because there were people vulnerable to indictment and other people prone to indict them.

SPATIAL DISTRIBUTION OF WITCH PERSECUTION

The location of witch trials in Early Modern Europe is shown in Figure 11. The map reveals that a large proportion of trials were con-

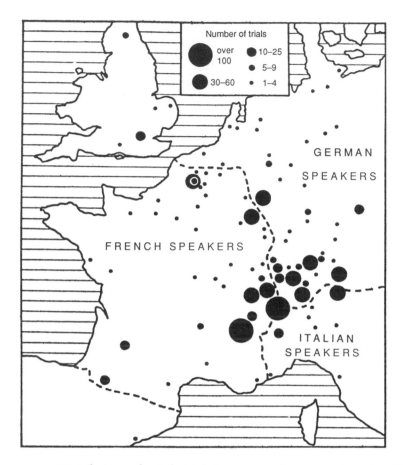

FIGURE 11. Distribution of witch trials in western Europe, 1580–1650. Reprinted from E. William Monter, "French and Italian witchcraft," *History Today* 30 (1980), pp. 31–35, by permission of *History Today*.

centrated in alpine areas of France and central Europe and in the Rhine Valley. In all these areas rye was the staple cereal. Temperatures in the coldest month (usually January) were below freezing, and at higher altitudes they might be colder. Witch trials were also more common in wet areas. The area of highest concentration of trials was both cold and wet. The large circle in the middle Rhône Valley (southeastern France) and the two circles in southwestern France encompass areas in which summer temperature tends to be ideal for production of ergot alkaloids (17.4°–18.9° C).[2]

Some other examples of the significance of spatial distribution may be found in the British Isles. In Scotland, witch persecution was concentrated in the northeast, along the coast—the country's main rye-growing area (Figure 12).[3]

The absence of persecution is also significant. There were few witch trials in Ireland, for example. One was held in Kilkenny in 1578, but the record is lost. In 1661 in the Puritan (English) colony of Youghal, there were victims of the bewitchment syndrome and a trial followed; likewise on Island Magee off County Antrim there was a trial in 1711. It may be relevant that the Irish at this time consumed mainly dairy products and oats, which may explain why they were not very susceptible to "bewitchment."[4]

In Spain, witch trials occurred only in Galicia, a damp region in

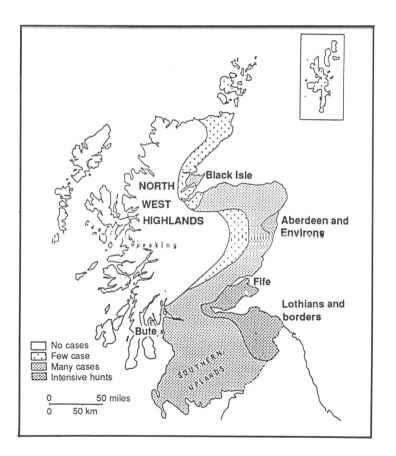

FIGURE 12. Distribution of witch trials in Scotland. Reprinted from Christina Larner, *Enemies of God: The witch-hunt in Scotland* (Baltimore: Johns Hopkins University Press, 1981), by permission of Johns Hopkins University Press, Chatto & Windus, and the Estate of Christina Larner.

the northwest corner, and in the Pyrenees.[5] Trials were rare in Italy and Scandinavia as well—countries that were warm and dry or that were in the cold northern tier.

TEMPORAL DISTRIBUTION OF WITCH PERSECUTION

It is generally agreed that between the end of the fifteenth century and 1560 the incidence of witch trials was low. There were some trials in the Pyrenees and Barcelona in 1507, 1517, and the 1520s. Epidemics of central nervous system disorders, not involving charges of witchcraft, were also rare. This was a period of warm weather.

In the 1560s the climate of Europe grew cooler and wetter. In Essex, France, Galicia, Switzerland, and southwestern Germany the incidence of trials increased.[6]

The best place for statistical testing of the relationship between witch persecution and ergotism is Swabia, in southwestern Germany. Erik Midelfort, using primary sources, compiled an annual index of the number of trials from 1550 to 1689. There is also an index of rye prices in Augsburg, which indicates the amount of pressure on the poor to eat grain of questionable quality. Moreover, other researchers have compiled an index of tree-ring widths in a nearby part of Switzerland that gives an indication of spring and summer temperatures during the period of persecution.[7]

I discovered that the higher the rye prices (at lag 1) in Augsburg, the more the witch persecution in Swabia ($r = 0.745$). The colder the spring and summer temperatures in Switzerland, the more the persecution ($r = -0.796$). The multiple regression coefficient r^2 equals 0.633 ($F = 20.7$), indicating that *63.3 percent of the variance in witch persecution could be predicted by temperature and price combined.*

The timing of witch persecution in eastern Europe was quite different. In Russia, the highest incidence of trials occurred in the period 1650–1700, and there were no trials before 1623. The mean winter temperature index for Russia in 1640–1749 (at longitude 35° E) was -11.1, compared to the 1100–1969 mean of -4.0, which indicates that cold winters predicted witch trials.[8]

In Poland, 81 percent of all witch trials occurred between 1700 and 1750.[9] Simultaneously, Germany, Sweden, the Baltic countries, and Russia, which, like Poland, were rye-dependent, were reporting epidemics of ergotism. These epidemics were distributed as follows:

in 1650–1699 there were four; in 1700–1750, thirteen; and in 1751–1800, eight.[10] The peak in symptoms of "bewitchment" in Poland and the peak in symptoms of ergotism reported by the neighbors of the Poles coincide. I suggest this is so because they had a common etiology.

WITCH PERSECUTION IN ESSEX COUNTY, ENGLAND

Many scholars have investigated witch persecution in England. G. L. Kittredge surveyed all published sources; K. V. Thomas studied the context of folk belief; and C. L. Ewen made an overall survey of the English court records.[11] Alan MacFarlane did an exhaustive study of the court records in Essex and evaluated various theories of witch persecution.[12] Essex was the only county in England with quarter-session records that went back as early as 1556; the next earliest series went back only to 1589, after the peak in witch persecution had passed; only four other counties had any records for the sixteenth century; and there were none surviving for sixteenth-century London.

MacFarlane showed that there was no evidence of a pagan cult in Essex, with fluctuating levels of activity, that might explain the incidence of witch persecution there. He found that Puritans were no more likely to persecute witches than were non-Puritans. Neither alleged witches nor their intended victims had any characteristic religious beliefs.

His examination of possible economic influences on the incidence of witch persecution yielded ambiguous results. On the one hand, he thought alleged witches tended to be poorer than their victims, although he gave no exact figures. In many cases an accusation of witchcraft was made after the "witch" had asked the "victim" for an economic favor; the victim having refused, the witch wished him ill. Moreover, in the period 1560–1650 the traditional means of helping the poor were "strained," according to MacFarlane. But he could find no correlation in Essex between the incidence of witch persecution on the one hand and, on the other, changes in population density, wheat prices, or the fortunes of the cloth industry. The central region of Essex, which was least troubled by poverty, had a high incidence of witch persecution.

MacFarlane rejected the notion that the incidence of witch persecution could be explained by an increasing incidence of illness, physical or mental. In three sample villages he found no correlation between mortality and witch accusations. Unfortunately, he had no morbidity statistics to consult. In his mind there were no connections between witch accusations and any particular disease. But MacFarlane, like most English scholars, did not consider that ergotism might have been present in Essex. He did not question the assertions of earlier authors, notably Charles Creighton, J. C. Drummond, and A. Wilbraham, that ergotism was insignificant in England and that little rye was consumed.

One reason Essex may have been the site of particular excitement over witches was the fact that it was a rye-growing center in a time of expanding markets, both within the county and in London. In order to expand production the farmers of Essex reclaimed marshland. According to Eric Kerridge,[13] reclamation took on greater impetus after 1560. This was particularly true of the Vale of London, in southwestern Essex, and along the Essex coast. One of the crops sown on recently drained marshland was rye, which flourished in the sour soil as other cereals could not. Moreover, this is the part of England with the highest mean July temperature (17° C in the twentieth century), the most favorable for ergot alkaloid production.

Are there traces of the symptoms of ergot poisoning in the Essex court records? Yes, and some symptoms were specific to severe ergotism:

(1) Permanent constrictures: a boy whose feet became "crooked and useless"; a child's hand "turned where the backes shoulde bee, and the backe in place of the palmes"; a man with "mouth drawne awrye, well neere uppe to the upper parte of his cheeke"; and a case in which "the right arme tirynge clene contrarie, and the legg contrarie to that, and rysinge double to the hed of the childe."

(2) "Fits": in three reports victims suffered violent "fits," which in common usage of the time usually meant symptoms of the central nervous system.

(3) Gangrene: one person's right thigh reportedly "did rot off."

These reports represented only a small proportion of all injuries attributed to witchcraft. Most of the other reported symptoms could have been symptoms of ergotism, but they neither proved nor disproved the case in themselves: they merely contributed to the larger picture. These symptoms included lameness; loss of sight, hearing, and speech; nausea; fainting; making animal noises; hallucinations ("a dazzling" in the eyes); suicide; and sudden death.[14]

Similar symptoms were noted in scattered court records and narratives in various parts of England where witches were persecuted. C. L. Ewen said that the greatest stir in witchcraft cases was caused by "the convulsive, hysterical, and epileptic seizures popularly termed 'fits.'" Michael MacDonald, in a recent study, noted that Richard Napier (1597–1634) regarded as bewitched patients who had plucking sensations, convulsions, "fits," and lingering symptoms generally.[15]

We have details of three case studies of bewitchment in England, none of which, unfortunately, occurred in Essex. They do serve, however, to make clear what English people in the sixteenth and seventeenth centuries meant when they spoke of bewitchment. The first was the case of Warboys village in Huntingdonshire. Here, in 1589, Robert Throckmorton's five daughters, seven maidservants, and others in the village suffered from fits, hallucinations, and temporary blindness, deafness, and numbness. In addition, many cattle in the area died.[16] The second was the case of Joan Harvey of Hockham, Norfolk, in the year 1600. Her symptoms included fits, spasms, the sensation of being "nipped," bloody spots on her skin, temporary blindness, deafness, and numbness, lameness, manic behavior, lying senseless, and hallucinations.[17] Norfolk, like Essex, was an important rye-growing center. The third case was an epidemic in Fewstone, Norfolk, which started in October 1621. The common symptoms included fits, hallucinations, and trances. One victim of alleged bewitchment had gangrene, and another threatened to commit murder and suicide.[18] The symptoms noted in case histories are typical of dystonic ergotism.

Generally, the victims of bewitchment were a generation younger than the accused witches. As noted before, ergotism affects children and teenagers most often and most severely. In a rapidly growing population such as that of Essex in the late sixteenth century there are more young people; as a result, the at-risk population for ergotism is greater than in a slow-growing, stable, or declining population. In 1584 Reginald Scot, speaking of England generally, said that witchcraft accusations were linked with the occurrence of "apoplexies, epilepsies, convulsions, hot fevers, worms, etc." in children.[19] According to MacFarlane,

> While suspected witches were characteristically middle-aged or old, their victims appear to have been younger adults. The Assize indictments often stated, in the case of children, the age of the victim; in a number of instances the victim was said to be the "son of" or "daughter of" another person. It seems likely that

this was only recorded when the victim was a child. . . . As well as these cases, there were sixty in which the victim was described as the "son or "daughter" of another. Comparing these ninety-two victims with a total of 341 victims altogether, it would seem that over two-thirds of those believed to be bewitched were adults. . . . Unfortunately it has been impossible to collect information on the exact age of the accusers; only indirect evidence, such as the presence of young children in the family of the bewitched, remains. This gives the impression that they were quite often a generation younger than the accused.[20]

That the victims of both bewitchment and ergotism were among the youngest members of the community supports the hypothesis that ergotism was a cause of the kind of behavior that was blamed on witchcraft.

Another piece of epidemiological evidence for this conclusion is the time of year of accusation. Indictments for witchcraft in Essex were about evenly distributed throughout the year, but were somewhat more frequent between February and June.[21] This pattern differed from the classic pattern of an ergotism epidemic identified on continental Europe, in which the disease peaked in August and September, immediately after a rye harvest. This is because the continental epidemics occurred in communities heavily dependent on rye alone; the inhabitants began to eat the new crop as soon as it was harvested. The English people, however, had a more varied diet; in time of scarcity they could turn to barley, oats, and legumes.[22] Moreover, judging from the practices of English settlers in New England, rye might stay in the barns until February or later before it was threshed and consumed.[23] Ergot can remain chemically stable for up to eighteen months.

Also, witch persecution was less common in certain sections of Essex than in others, and these areas were at less risk for ergotism. Relatively little rye was grown in the northwest corner, a region of chalky hills, and in this area persecution was rare. Another part of Essex with relatively little persecution was around Colchester: little rye was grown there. The southeastern coastal area of Essex, bordering on the Thames estuary, was almost free of witch persecution: wheat was the most important crop in this area. But there was a high incidence of witch persecution in the southwestern section (Vale of London) and northeastern section of Essex—lowland areas with wet, sandy soils, where ergot was more likely to form on rye and where rye was cultivated intensely.[24]

The incidence of witchcraft accusations was related, in addition, to long-term trends. Witch trials in Essex fell mainly between 1560 and 1619, with a concentration in the 1580s and 1590s. During this period the best predictor of trials was wet growing seasons ($r = 0.559$), but changes in diet in the later years may have had an even greater effect. With the growth of the London market in the late sixteenth century, some farmers prospered (and were thus able to afford white wheat bread) and some lagged behind. The poor ate oats, barley, and other cereals, but could not afford either wheat or rye. According to William Harrison, writing in 1587 after cereal prices had risen, "If the world lasts awhile after this rate, wheat and rye will be no grain for poor men to feed on, and some caterpillars [pillagers] there are can say so much already."[25] According to Kerridge, around 1620 the people of southeastern England had largely lost the habit of eating rye.[26] In Scotland, however, where rye consumption continued during the seventeenth century,[27] ergot symptoms and witchcraft accusations continued.

THE SOCIAL RESPONSE

The distribution of witch accusers and persecutors depended on the social structure of the region involved, and no European-wide generalizations can be made. The distribution of the *accused*, on the other hand, was nonrandom and tended to be the same in all rural areas.

Witches had wortcunning, knowledge of the medicinal properties of herbs. The role of the healer was a perilous one, for people were afraid of his or her seemingly magic power over a living body. They might think that someone who could cure disease by magic could also cause it by magic. Restoring health was "white" magic, taking it away was "black" magic. Like any physician today, a witch could be blamed when things went wrong, and in some ways witchcraft accusations may have been analogous to malpractice suits against physicians.

Among the symptoms healers of the Early Modern Era could sometimes relieve were the very symptoms associated with both ergotism and bewitchment. Motherwort and mistletoe, for example, were effective against some kinds of convulsions and spasms.[28] It is not unlikely that, where ergotism was mistaken for bewitchment, the targets of witch-hunts were often among those herbalists who had had some success in quieting nervous disorders. On the other hand, witches could not be accused of causing bubonic plague and were not so accused, for they had no means to cure it. The same was true of other diseases for which they had no cure. Witches were usually accused of causing diseases which, in some cases, they knew how to cure or relieve.

Accused witches might admit to curing disease while denying that they caused it or that they used magic to cure it. One accused Essex witch, Ursula Kemp, declared that though she could "unwitch," she could not "witch."[29] In Scotland in 1620, Alexander Drummond was accused of using magic to cure "frenzies, the falling evil [epilepsy], persons mad, distracted or possest with fearful apparitions, and St. Antonie's fire [gangrenous ergotism]." He admitted curing people, but said he used no magic.[30]

In summary, in Early Modern Europe witchcraft persecution occurred at a time of widespread impairment of the health of people and animals. The distribution of illness, often interpreted as a sign of bewitchment, mimics the pattern of the incidence of ergotism: it was most common in alpine areas and those with summers in the 17.4°–18.9° C temperature range; a majority of the victims were children and teenagers; and rye was a dietary staple in the areas affected.

Why did witchcraft persecution peak in the period 1560–1660? Perhaps the weather was to blame. This was a cold century. The Thames River froze over in 1565, 1595, 1608, 1621, 1635, 1649, and 1655; it has not done so since.[31] Cold winters traumatize rye and increase the risk of ergot alkaloid formation. Such alkaloids may have caused the symptoms of "bewitchment." When the incidence of these symptoms increased, so did the incidence of witchcraft persecution. We today should avoid the mistake made by the witch-burners of long ago by not overlooking a physical cause for events that mystify us.

CHAPTER SIX

The Great Fear of 1789

BETWEEN July 20 and August 6, 1789, waves of panic swept over the French countryside. The new rye crop was just harvested, but there were rumors that brigands were coming to seize it. Many people believed they had glimpsed these bandits and feared it was already too late: women would be raped and murdered, children massacred, homes set afire. As tocsins rang, the peasants, weeping and shouting, fled into the woods to hide or armed themselves with pitchforks, scythes, and hunting rifles.

In Dauphiné, a region of France in the upper Rhône Valley, peasants looted and burned châteux, but this was an isolated "political" incident: most peasant behavior was disorganized; *la Grande Peur* was not a rising of the masses. When the panic was over, some peasants blamed the rich, calling it a trick to deprive them of their daily pay.[1]

This brief episode might have been forgotten had it not been part of the French Revolution, by which we date the end of absolute rule of the privileged classes in France and the emergence of one of the first strongholds of the idea of popular sovereignty. The "Great Fear" among the landless aroused a great fear among landowners, an apprehension that the peasants might seize property and turn upon their masters. To forestall any such catastrophe, the National Constituent Assembly met on the night of August 4 in Versailles, in what one contemporary observer called a strange state of "patriotic drunkenness," and voted to abolish what was thereafter to be called the *ancien régime*. Fearing further peasant disorders, the nobles renounced their seigneurial dues, others gave up the clerical right to collect tithes, and the Assembly ended the sale of public offices and privileges. King Louis XVI took no part in this affair, but the assembled delegates proclaimed him "the man who restored liberty to France."[2]

In recounting the events of those eighteen days in 1789, historians often record one very puzzling fact: the peasants' fears were exaggerated. At the time of the Great Fear, vagrants were roaming the countryside in search of food, but they were apparently neither organized nor dangerous. Passing over this inconvenient fact, some scholars explain the Great Fear as an insurrection of peasants who resented paying taxes and tithes.[3] In the spring of 1789 there were peasant protests against the food shortage and "feudal" practices, but the Great Fear of July and August was mainly a *panic*, not a protest.

Presenting a detailed chronology and a map illustrating the spread of the panic in his study, *The Great Fear of 1789*, Georges Lefebvre argued convincingly that there had been no conspiracy among the peasants. The panic did not spread from house to house but affected whole communities simultaneously. Undaunted by obstacles that would stop a human traveler, it appeared even to scale mountain peaks without difficulty. Citing local documents, Lefebvre stated that outside the Dauphiné the panic was not an expression of resentment against injustice. Rather, the country people were terrified. Although Lefebvre reconstructed the mental state of the peasants, he could not account for either the location or the timing of events— why the panic had occurred in some parts of France and not in others, in the summer of 1789 and not at some other time.

There is an explanation for these puzzling facts, although historians have generally overlooked it. The clues are buried in eighteenth-century French provincial records, which show that in the summer of 1789 many French citizens may have suffered from ergot poisoning. These same records also mention that in the region of Artois and in the Maisonnais parish, in the commune of Champniers-et-Reillac, many women miscarried in these summer months.[4] And in the *Histoire et mémoires de la Société Royale de Médecine* of the same year, one Dr. Geoffrey chronicled a marked deterioration in public health in the second half of July, reporting that jaundice, diarrhea, and nervous attacks were common, especially among women. Within the space of two weeks, he said, he had seen five patients who had "lost their heads"—become manic or imbecile or appeared dazed. Later, in August and September, he found many cases of stomach pain, diarrhea, and colic. Geoffrey attributed all of the symptoms to the consumption of "bad flour" and reported that all were relieved by a change to "better bread."[5] Two Paris physicians also chronicled an increase in illness, especially nervous diseases, in the second half of July. When their patients, many of them pregnant women, suffered "apoplexies, paralysis, anxiety, fear, visceral upset, depression, slow fevers, and erysipelas," these doctors, like Geoffrey, suspected that "bad bread" might be to blame.[6]

An especially promising clue to the mystery of the Great Fear appears in a 1974 study of public health in Brittany by Jean-Pierre Goubert. Goubert quotes a physician living in the town of Clisson as saying that in July 1789 the rye crop was "prodigiously" affected by ergot. Although Goubert drew no inferences about the Great Fear, he reported that toxic ergot horns were found on about one-twelfth of all ears of rye.[7] This was indeed a prodigious amount; rye flour containing one percent ergot is sufficient to cause a full-blown epidemic.

Physicians were slow to recognize that contaminated rye was dangerous. In 1602, to be sure, a French physician recommended that nursing women eat white bread in order to keep their babies from having spasms.[8] Several eighteenth-century observers described epidemics of what we now know to be dystonic ergotism, mentioning symptoms of "delirium" but failing to identify rye as the cause.[9] In 1758 Joseph Raulin, a Paris physician, wrote about a disease, which he called "vapors," whose victims manifested all the symptoms of dystonic ergotism. "Clouds of different colors," he said, "appear to the eyes like a rainbow"—that is, the victims hallucinated—but Raulin, too, missed the connection with contaminated rye.[10] Other learned men of the 1770s linked ergot with gangrene and spasms but overlooked its power to produce mental disorders.[11] It was not until 1785, when Johann Taube's study of the dystonic ergotism epidemic of 1770–1771 was reviewed in France, that any French physicians recognized this danger.[12]

A Russian study of psychic disturbances associated with ergotism, published in 1893, revealed that unpleasant hallucinations and panic were more common than pleasant hallucinations. The victims "believe themselves to be drowning, or they see a fire and fear to be burned alive. Again others believe that someone is attacking them in order to butcher or strangle them. Some see robbers attacking their homes and others see devils, demanding they give up their faith."[13]

One might wonder why so many peasants ate bad bread in 1789. First of all, they ate a great deal of bread *every* year—two or three pounds of it a day when they could afford it. The average daily intake among the poor was only between seventeen hundred and two thousand calories, of which 95 percent came from cereals.[14] Moreover, 1789 was an "ergot year" in northern France. Since 1697, when reasonably complete records were first kept, France had not seen weather conditions so favorable to the growth of ergot on rye. The unusually cold and snowy winter of 1788–1789 weakened the rye, and the cold and humid spring allowed fungus to grow on the plants. In early May a dry, warm period set in, promoting the spread of spores by wind from one plant to another, followed by a warm, wet summer ideal for alkaloid formation. In Paris it rained twenty out of thirty days in June.[15] There had been a disastrous crop failure in 1788, and the hungry peasants probably consumed the newly harvested rye of 1789 hurriedly, without carefully cleaning it. Immediately after harvest the ergot was at maximum toxicity.

No physician, then or since, has suspected that ergot in rye was to blame for the peasants' visions and panic on the eve of the French Revolution, but the sequence of panic outbreaks in July and August could be accounted for as follows: (1) Ergot was more likely to form on rye in the moist climate of northern France. Hence the five centers of panic diffusion were in the north (see Figure 13). (2) Yields of rye were higher in northern than in southern France, so in a year of dearth, surpluses were sent from the north to southern markets. Panic symptoms appeared later in the south than in the north.

In Figure 13 the distribution of panic symptoms is imposed on that of rye and maslin (a mixture of rye and other grains) production. The map is based on yields in the later 1830s, the earliest available systematic data on crop distribution, corrected to represent production in 1789 on the basis of both contemporary writings and modern studies of the area.[16]

Some areas in which rye and maslin were grown were exempt from panic but this may be explained. Most of the rye grown in 1789 in Vannes (Brittany) (letter A in Figure 13) was used to pay taxes and exported: the commonfolk ate mainly buckwheat. When the rye was locally consumed, it was mixed with oats and barley. In 1789, on account of excessive summer rains, most crop harvests were poor: wheat was one-half normal; rye, seven-twelfths; oats, two-thirds. Only the buckwheat harvest was normal. So rye consumption in the summer of 1789 in Brittany was probably insignificant.[17] This may also have been true of the area north of Nantes. The cultivation of rye in the *landes* of Bordeaux (letter B) was mainly a development after 1789.[18] In the Sologne (letter C) cultivated rye was carefully

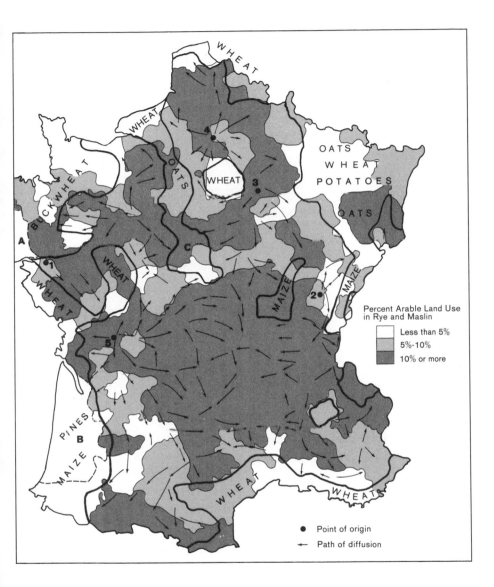

FIGURE 13. Distribution of episodes of panic, July 20–August 6, 1789, and production of rye and maslin, as percentage of acreage per arrondissement, 1836–1838, corrected to approximate acreage of 1789. Numbers indicate what Lefebvre thought were the centers of diffusion of rumors. For sources see notes 17–21 of this chapter.

cleaned. Ergot here normally caused gangrenous rather than dystonic ergotism.[19] In Roussillon (in the eastern Pyrenees) disturbances occurred, but later, in September and October 1789.[20]

Other rye-growing areas that were exempt from panic were areas in which alternative starches were more abundant: buckwheat, maize, wheat, oats, and potatoes (see Figure 13). The availability of these foods meant that the peasants did not have to eat the rye crop as soon as it was harvested, before proper cleaning, when ergot toxicity was maximal. Moreover, fungal infections of crops are less severe in areas in which several crops are grown instead of one.

In Normandy (and probably elsewhere) the areas affected by the Great Fear were not only areas in which the poor ate rye bread, but also areas in which hallucinations and spasms were endemic.[21]

Unfortunately I have not had the opportunity to sift archival data in search of precise sequences of harvesting rye → consumption of rye → appearance of panic. I have been presenting a series of interesting coincidences that make a causal connection plausible.

Panic was not the only symptom to appear: during 1789 and 1790 other unusual mental states were widespread in France. In Grenoble a group of "convulsionaries" made a stir. In keeping with apocalyptic beliefs of the time, they were convinced that the return of the Jews was imminent, that "Elias has come, that he is getting ready to carry out his mission very soon," and that the "reign of a thousand years of Jesus Christ is at the point of beginning." In Périgord, after the Great Fear, the prophetess Suzette Labrousse began to gain a following. As Clarke Garrett declared, "in 1789 and 1790, it was widely believed in France that religion and revolution would triumph together."[22] It may be recalled that ergot retains its toxicity for up to eighteen months; perhaps some of these later examples of mass hysteria are simply delayed effects of the blight of ergot of 1789.

If the Great Fear of 1789 was a manifestation of ergotism in France, one would expect this disease to appear in the neighboring countries where rye was also a staple food at this time. Some rye was grown in northwestern Italy, and in 1789 an ergotism epidemic was reported in a conservatory for young ladies in Turin, a city where mortality was 43.4 percent above normal that year.[23] Rye was still a staple food in parts of the Midlands and in northern England, and in Manchester and Selford (in the wettest part of England) symptoms of ergotism were reported in 1789.[24] James Jenkins, a Quaker merchant, observed that 1789 was "a time of strange delusions in which Friends were too much concerned—I was myself one of very many who resorted to Richard Brothers, the Prophet."[25] In England that year, however, spring rainfall was only average and the mean summer temperature only 15.3° C, too cool for ergot alkaloid production.[26]

In Germany, a country of high rye consumption, the symptoms of ergotism were not reported. The mean July temperature in Berlin was 19.0° C—too high for ergot alkaloid production; in Vienna, it was 21.4° C; and in Vilna, 20.0° C. Moreover, after the widespread ergotism epidemic of 1770–1771, the Germans learned how to diagnose the disease and local governments took steps to prevent the consumption of toxic grain.[27]

In summary, then, it appears likely that certain ergot alkaloids created a suggestible state of mind among rye-eating populations in several European nations in 1789 and 1790. Cultural factors determined the precise nature of the interpretations that victims placed upon their symptoms. In some cases, the suggestible mental state manifested itself as visions of brigands coming to steal crops; in others, as apparitions of the millennium. Cultural factors, varying by time and place, gave these visions shape and color and conditioned reactions of fear or of religious fervor.

The mental symptoms of ergot poisoning may be dramatic, but in most cases they are transitory. In the next chapter I will show how undramatic dietary changes reduced the incidence of mold poisoning and reshaped the population history of Europe. The Great Fear of 1789 was among the last of the bizarre events of European history that may be explained by food poisoning.

CHAPTER SEVEN

The Population Explosion of 1750–1850

Beginning around 1750 Europe's population began to balloon. For a hundred years it grew with unprecedented steadiness, 0.5 percent to 1.5 percent annually.[1] From the level of 140 million in 1750, it reached 250 million in 1850—a gain of 78.6 percent in a century (see Table 8).

The swelling population had important consequences for the In-

TABLE 8. Population growth in selected countries of Europe, 1750–1850 (in millions)

Country	1750	1800	1850	1850 as percent of 1750
England, Wales	6.00	9.25	18.00	300.0
Scotland	1.25	1.50	3.00	240.0
Ireland	3.00	5.25	6.50	216.7
Scandinavia	3.75	5.25	8.00	213.3
France	24.00	29.00	36.00	150.0
Belgium/Luxembourg	2.25	3.25	4.50	200.0
Netherlands	2.00	2.00	3.00	150.0
Germany	15.00	18.00	27.00	180.0
Poland	7.00	9.00	13.00	185.7
European Russia	26.00	36.00	60.00	230.8
Switzerland	1.50	1.75	2.50	166.7
Rumania	3.50	5.50	8.00	228.6
Spain	8.00	9.50	11.50	143.8
Italy	15.00	19.00	25.00	166.7
Total	118.25	154.25	226.00	191.1

Source: Colin McEvedy and Richard Jones, *Atlas of world population history* (Harmondsworth, 1978), pp. 19–109.

dustrial Revolution, levels of political violence, and the flow of im-
migration to the New World. It would be difficult to find any aspect
of human experience that it did not touch in some way. How did it
all begin? What sustained this trend?

MARRIAGE HABITS, THE POTATO THEORY OF POPULATION, AND OTHER POSSIBILITIES

For a more precise definition of the problem there are available, as a
result of the meticulous work of E. A. Wrigley and R. S. Schofield,
some new estimates of the changing size of the English population
and its fertility, nuptiality, and mortality rates.[2] Wrigley and Schofield offer as a tentative explanation for population growth after 1750
the earlier first marriage of females. Early marriage, they say, could
account for more than half the increase in the gross reproduction rate
of England between the second half of the seventeenth and the first
half of the nineteenth century. They suggest that a decline in the
mean marriage age of females from 26.5 years (1650–1699) to 23.4
years (1800–1840) would increase fertility by about a quarter. Also,
more young people may have married in the 1800s than in earlier
times. Pointing out that population growth and economic development were positively correlated after about 1750, they argue that the
arrows of causality should run from economic growth to earlier and
more universal marriage, and thence to increased fertility.

We owe much to Wrigley and Schofield, and we can rejoice in the
completion of their enterprise. Unfortunately, their explanations are
less than satisfying. Several scholars have pointed out some of the
deficiencies, among them:[3]

(1) Wrigley and Schofield may have mistakenly used their data to
obtain high fertility levels for 1800–1820.

(2) Their model cannot explain differences in patterns of fertility,
nuptiality, and mortality between England and France.

(3) Early industrial development does not necessarily stimulate
earlier marriage. For example, when entrepreneurs in the late eighteenth and early nineteenth centuries installed spinning machines in
the valleys and tributaries of the Lower Seine, despite the new economic opportunity couples in Normandy still married late: 27.0 years
for women, 28.4 years for men.[4]

(4) There is no necessary connection between a fall in the mean
age of females at first marriage and an increase in the number of
children who are conceived and who survive long enough to repro-

duce.[5] A rise in conceptions might be canceled by an increase in maternal deaths, miscarriages, stillbirths, and neonatal mortality—unless some other factor intervened. Teenage pregnant women are particularly prone to certain diseases (for example, tuberculosis) and to interrupted pregnancies. The "telescoping" effect of generations produced by earlier marriage will occur only if these negative factors are kept under control.

(5) Wrigley and Schofield did not consider the possibility that there might be substances in nature capable of suppressing fertility and crippling the human immune response.

There is no proof that English population increased because of change in its marrying habits. Nor is there any proof that the Industrial Revolution caused population increase: the reverse is more likely. But Wrigley and Schofield have served up a banquet of unexplained, interesting facts that can sustain quite different inferences. For example, the last mortality crisis with 25 percent or more deaths above trend occurred in 1741–1742. After 1750 infant and child mortality declined among commoners and even more rapidly among peers.[6] Significant changes took place in the seasonality of conceptions and of crisis mortality after 1750. Mortality crises tended to follow severe winters, and mortality varied markedly from one region to another and even from one parish to another. Wrigley and Schofield have little to say about these facts of death.

Even if economic development did serve to stimulate English population growth, it could not have been responsible for the patterns of population growth that occurred elsewhere in Europe. The population of Ireland began to explode in the early eighteenth century. The population of Norway, Denmark, and Sweden increased rapidly in the late eighteenth and early nineteenth centuries, before any industrial development occurred.[7] In France fertility declined after 1770, mortality after 1795. The explanation for English population growth suggested by Wrigley and Schofield, even if true, is not generalizable.

It therefore seems justified to continue to seek the causation of population growth in Europe along different lines. How can one explain the prolonged mortality decline after 1750? It now seems reasonably clear that improvements in sanitation and medical care, the decline in war casualties and deaths associated with famine, or even smallpox inoculation cannot be taken seriously as solutions. For example, James Riley believes that better water and sewage systems reduced deaths from diseases transmitted by insects, but his best evidence, the leveling of the late-summer mortality peak, may have another explanation.[8]

I believe that Thomas McKeown and Jean-Noël Biraben have pointed in the right direction: to changes in nutrition.[9] An improvement in any of three dimensions of nutrition—quantity of calories, balance of nutrients, and the absence of harmful substances in food—may serve to reduce human mortality.

Since at least 1843 the introduction of the potato into the diet of northwestern Europe has been proposed as the answer to the puzzle of population growth.[10] The potato gives about four times as many calories per acre of land as does wheat. Land planted in potatoes can support four times as many people as the same land planted in wheat. A diet of potatoes and milk, upon which the Irish poor subsisted in the eighteenth and early nineteenth centuries, is highly nutritious. And so, the argument has gone, because the potato provided the calories and nutrients to sustain a larger population, a larger population came into being.

The potato theory of population growth has not been wholly blighted by persistent criticism. To grow potatoes one must plant seed potatoes, and a seed potato contains more calories than a seed of wheat; the yield of potatoes per calories planted is about the same as the yield of wheat per calorie of wheat planted. One is not growing more food per unit capita, but the same amount of food on less land with less labor. The switch is equivalent to miniaturizing agricultural production. The trouble is that this increased efficiency does not explain health improvement. In terms of the first two dimensions of nutrition—the number of calories and the balance of nutrients— the English were better fed in the first half of the eighteenth century than the Irish. Yet the English population grew more slowly than the Irish.

The potato theory may contain the right answer for the wrong reasons. If it is amended in a crucial detail, it may be more plausible. An important cause of high mortality in England may have been negative nutrition: namely, the presence of mycotoxins in cereals. If potatoes replaced cereals, what would happen?

A potato attacked by fungi tastes and smells rotten and is obviously unhealthy to eat, so it is discarded. Normally nutritious, potatoes signal their own corruption. But microfungi that infect cereals may do so insidiously. Their presence may be hard to detect if one is not on guard, since the infected cereal is not unpleasant to smell or taste. Such cereals are likely to be consumed.

There is another advantage to potatoes if the cereal they replace is rye. Rye may contain fertility-suppressing alkaloids as well as lethal mycotoxins. If potatoes were available in time of dearth, the poor would not have to eat ergoty rye and fertility would not decline.

What was true for England may be highly generalizable. After 1750 population growth occurred throughout Europe and in the American colonies and China. Dietary changes were not the only influence at work, of course. Between 1742 and 1768 growing conditions were unusually warm in Europe and discouraged the activity of *Claviceps purpurea* (ergot).[11] In other words, nutrition may have improved because of climatic, not dietary, change.

Far less is known about health on the European continent than about health in England, but it is enlightening to compare the English with the French, who were almost equally interested in collecting statistics about themselves. I will not compare the two populations as to fertility, for that is a problem that involves some factors beyond the scope of this book. What follows is a comparison of differing mortality trends in the two nations.

MORTALITY TRENDS IN ENGLAND

According to Wrigley and Schofield, the crude death rate for England, excluding London, gradually declined after 1750, from 29.2 per thousand population in 1700–1749, to 27.0 per thousand in 1750–1799, to 24.0 in 1800–1849. What changes in diet were occurring? The substitution of wheat for rye had begun early: by the 1620s the common folk in much of southeastern England were no longer eating rye bread.[12] Consumption of rye remained significant in the Midlands and in the north, but after 1682 wheat prices tended to fall and wheat consumption presumably increased. In 1751 Charles Deering wrote of the people of Nottingham (in the Midlands): "Even a common washerwoman thinks she has not had a proper breakfast without Tea and hot buttered White Bread."[13] Wheat consumption in the north of England increased more rapidly after 1770; by 1800 it reached 11 percent to 32 percent, depending on the area. It has been estimated that by 1800 about 70 percent of the population of England and Wales ate wheat habitually.[14] Meanwhile, after 1770 the poor began to eat more potatoes (see Figure 14).[15]

As rye consumption declined, conceptions in late summer and early fall became more frequent. As the poor resorted to potatoes

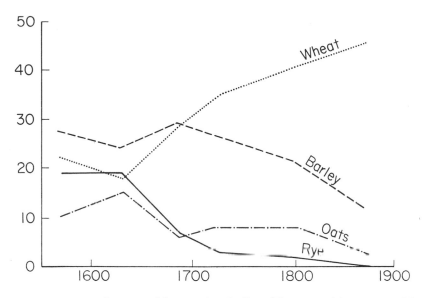

FIGURE 14. Production of four grains in East Worcestershire, 1540–1867, as percents of total population.

instead of moldy grain in time of dearth, mortality in late summer and early fall flattened out.

As noted above, England had no more mortality crises after 1741. Mortality never increased more than 25 percent over trend. There were only three times in which it rose by more than 15 percent:

(1) In 1762–1763 mortality rose by 24.2 percent. The winter of 1763 was severe and the following growing season, wet. Mold poisoning is a plausible explanation.

(2) In 1783–1784 English mortality was 16.7 percent above trend. The winter and spring of 1784 were cold and the summer was wet. On June 8, 1783, Mount Lakagigar in Iceland began a large eruption lasting to February 1784, and over this period it produced a large volume of tephra. A thick, dry haze, smelling of sulfur and hurting the eyes, spread over Europe. Crops in Scotland, Ireland, and the Netherlands were damaged in 1783–84. This disaster may have increased deaths involving mycotoxins.

(3) In 1846–1847 English mortality rose 15.1 percent above trend; the Scottish mortality level was also above average. An estimated one million Irish died in the Great Famine. The winter of 1845 was a cold one, and during the next summer, both cold and wet, the great potato blight began not only in Ireland but all over Europe. The summer of 1846 was warm and very wet, and the blight returned, com-

pletely destroying the crop. The crop of 1847 was sound but small, but the blight returned in 1848. It may be relevant that on September 2, 1845, Mount Hekla in Iceland erupted. Tephra was observed as far away as County Sligo, in northwestern Ireland, in 1846, where it damaged the leaves of the potato crop and marked other objects in the open with black spots like ink.[16]

But these three mortality crises were far less serious than previous crises. Heavy toxic fallout from unusual nearby volcanic eruptions may serve to explain two of them. Improving health conditions could not have been caused by declining volcanic fallout, because volcanic activity was not declining. Diet, on the other hand, was changing.

One objection to theories attributing mortality decline to better nutrition is the fact that before 1725 there was no significant gap between the mortality rates of the peerage and those of the masses. Because England's upper class ate white wheat bread, one would expect to find class differences in fertility and mortality. In Early Modern England, however, the children of aristocrats, usually cared for by servants and wet nurses, were exposed to ergot poisons in the breast milk of their wet nurses and to both ergot and *Fusarium* toxins in the solid food the servants gave them. Infants, especially, might ingest a lethal dose of toxins in a single meal and die suddenly of "convulsions."

Prior to the second quarter of the eighteenth century, infant and child mortality among both peers and commoners was high. From 1550 to 1725, between 16 and 24 percent of the children of peers died before the age of fifteen. From 1550 to 1749, between 21.2 and 27.0 percent of all commoner children died before the age of nine.[17] The children of peers were better off, as one would expect, but the gap was not very wide.

The pediatric literature of Early Modern England shows that the affluent, who read such literature, were ignorant of the dangers of mold poisoning. The doctors of Tudor and Stuart England listed as common symptoms for children "spasms," "palsy," "epilepsy," "erysipelas," "sacer ignis," and "St. Anthony's Fire"—all specific symptoms or names of ergotism. They also mentioned "quinsy," a swelling of the throat accompanied by a tendency to bleed and skin eruptions, which suggests an immune-deficient condition.[18]

In 1733 and again in 1742 George Cheyne urged a diet of milk, seed, grain, and vegetables for nervous disorders: he omitted rye from his list of recommended grains, but did not warn specifically against it. Significantly, he reported that a diet limited to milk could remove barrenness among some couples of great families who wanted heirs.[19]

In 1765 Simon-André Tissot published an account of ergotism and its cause in the leading English scientific journal, *Philosophical Transactions*.[20] By 1772 some doctors believed that harvest-time "agues" and "fevers" occurred after an unusually wet season and were associated with "unwholesome food from the spoiled grain."[21] And by 1781 it was observed that pap made from bread might cause fatal convulsions in infants.[22]

Moreover, eighteenth-century aristocratic mothers began to do something physicians had long recommended: to breast-feed their own infants and to involve themselves more directly in the care of their children. Those aristocratic women who were subject to depressions, "hysteric fits," and other nervous disorders were believed to be unsuitable as nurses, so wet nurses might still be used. But the wealthy tended to avoid employing wet nurses who had such "nervous" symptoms and to feed those whom they employed white bread.[23] For this reason, it may be suggested, mortality among peers declined, beginning in the period 1725–1749, and continued to do so up to 1849 (Table 9).

After 1750 the mean life expectancy at birth of peers began to

TABLE 9. Mortality of English peers' children, 1550–1849

Peer cohort born	Percent of children who died before age 15 in peer marriage of completed fertility
1550–74	16.00
1575–99	17.24
1600–24	20.72
1625–49	24.73
1650–74	20.80
1675–99	21.51
1700–24	21.79
1725–49	15.61
1750–74	13.85
1775–99	12.17
1800–24	11.34
1825–49	8.24

Source: T. H. Hollingsworth, "The demography of the British peerage," *Population Studies*, supplement, 18, no. 2 (1964), p. 42.

extend farther than that of commoners. During 1750–1774 a peer could look forward to 45.1 years of life, a commoner to only 35.7 years. The gap continued to widen: by 1825–1849 the life expectancy for peers was 55.3 years, and for commoners (1826–1846), 40.2 years. Differences between the two groups in infant and child mortality accounted for the gap.[24]

It should be noted that life expectancy for commoners began to rise during the last quarter of the eighteenth century; potato consumption increased at the same time; but no improvement in real wages occurred until about 1820.[25]

Clinical reports, too, indicate a decline in acute immune-deficient conditions. By 1782 James Hutchinson in Edinburgh noticed that the "fevers" of midcentury were becoming less common and suggested that increasing consumption of the potato and other dietary changes might be responsible. In 1786 James Sims of Tyrone, who left Ireland by 1772, said that the last epidemic of "scarlatina anginosa" (apparently an immune-deficient condition) had occurred there in the spring of 1755. From 1805 to 1831 symptoms of acute immune deficiency were no longer reported in England. They also became rare elsewhere in Europe, except in France.[26]

Mortality in London. When Wrigley and Schofield reconstructed the population history of England they omitted all London parishes. Not only did this give their work a rural bias, but it also left out of account the most complete evidence on causes of death in all of England.

Using the aggregate specific death ratios of the eight leading causes of death, and splicing the data of the parish Bills of Mortality with the later data of the General Registry, one can see that the period of greatest health improvement prior to the twentieth century was the period 1795–1834 (Figure 15). Especially sharp were the declines in deaths from "convulsions" and "consumption," which affected mainly infants and young adults, respectively (Figures 16 and 17).[27] Although the causes of these declines have not been determined, it is known that potato consumption was increasing rapidly, and rye consumption declining, during 1795–1834. A decline in food poisoning would tend to reduce deaths among infants and young adults, who are especially prone to it.

MORTALITY IN FRANCE

Considerable progress has been made in the study of overall population trends in eighteenth-century France. From 1700 to 1787 the

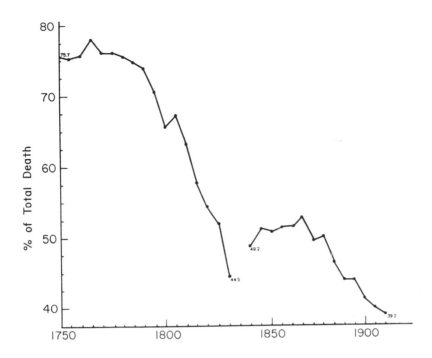

FIGURE 15. Sum of specific death ratios of the eight leading causes of death in London, 1750–1909.

French population increased by only 32 percent, while the English increaoed by 61 percent and the Irish by 110 percent.[28] Only in the last years of the eighteenth century did the common people in most parts of France shift from rye bread to white wheat bread, potatoes, and only then did population growth gain momentum. Before that, growth was unsteady and localized.

Population increase in Early Modern France is characterized by four phases. Growth was slow from 1670 to 1745, each gain being virtually wiped out by deaths from epidemics. From 1748 to 1778 the rate of growth increased; it was similar to that of England until about 1770. From 1779 through 1810 French population once again stabilized, the gains from good years being canceled by epidemics in bad years. When, after 1810, the population began to grow more rapidly, it was not because fertility increased but because mortality decreased.[29]

The seasonal pattern of mortality in France resembled that of England, with a peak in March–April and another peak in July–September.[30] This pattern remained stable in France from the sev-

FIGURE 16. Percent of deaths from "convulsions" in London, 1750–1909.

enteenth through the early nineteenth century. Around 1750 an interesting contrast appeared between the two countries: the English ate less rye, conceived more children and died less often during the summer months of July–September, and began to increase rapidly in number. The French, on the other hand, continued to eat as much rye after 1750, continued to conceive few children and to die at the same rate during July–September, and did not increase significantly in number.

From 1748 to 1778 health conditions improved in both countries, and it seems probable that the weather was responsible. During that period in no year were conditions ideal for the development of either ergot or *Fusarium* toxins, and there were only two unusually wet years (1751 and 1777). By contrast, in the preceding 33-year period (1715–1747) there were two years (1734 and 1739) ideal for the development of *Fusarium* toxins and six other unusually wet years (1715, 1727, 1728, 1732, 1744, and 1745). In warm Paris during 1757–1778 only two summers out of twenty-two were cool enough to favor ergot alkaloid formation. In cooler central England only four sum-

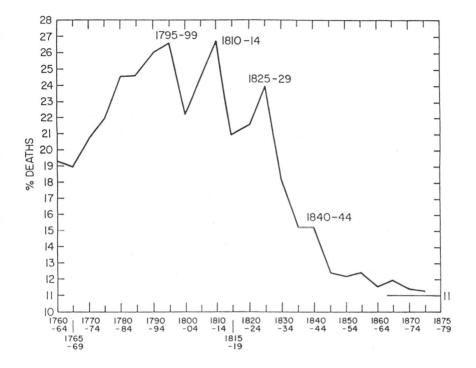

FIGURE 17. Percent of deaths from "consumption" in London, 1750–1909.

mers in this period were warm enough to favor ergot alkaloid formation.[31]

The course of public health in France diverged from that of England beginning in 1779. For the rest of the century French fertility was low and mortality high. Perhaps the weather, once again, was responsible. Between 1779 and 1799 there were two years (1789 and 1790) ideal for the development of ergot, and two other unusually wet years (1782 and 1792) were favorable to fungal growth in general. In the Paris region twelve out of twenty-one summers were cool enough for ergot alkaloid production, while in central England only five were warm enough, and just barely so.[32]

Among the various regions of France there were very large differences in population growth rates during the eighteenth century (see Figure 18). These were correlated with dietary differences. In Alsace-Lorraine, to the east, the population grew by 100–200 percent, compared with the national average of 28 percent for 1700–1787. Here in the east the potato was introduced early in the seventeenth century and was a major element in the diet by 1778. Infant and child mor-

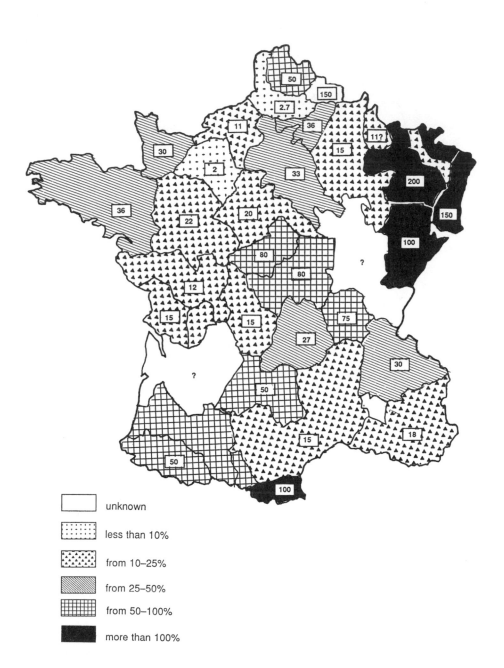

	unknown
	less than 10%
	from 10–25%
	from 25–50%
	from 50–100%
	more than 100%

FIGURE 18. Percent of population growth in different regions of France, 1700–1787. Reprinted from Michel Morineau, "Les Faux-semblants d'un démarrage économique," *Annales Economies, Sociétés, Civilisations: Cahiers des Annales* (1970), p. 296, by permission of *Annales*.

tality in the seventeenth and eighteenth centuries in the east was below average for France. The correlation between potatoes in the diet and population increase is not complete, however. In the Dauphiné, for example, where potatoes were important in the diet, population growth was only 15 percent, below the national average. The reason for this exception may be that people of the Dauphiné were also eating large amounts of rye bread. Child mortality was above average and marital fertility below average in this region.

Population increased in areas where grains besides rye were common or where conditions were not favorable for molds. An area of rapid population growth was Hainault-Cambresis, in the northeast, where wheat was an important crop. In the Perpignan area, in the eastern Pyrenees, the population doubled. Here people were adding potatoes to a maize and rye diet, and the danger of molds was less in the south because of the warm and arid climate. There were above-average increases (50–80 percent) in the southwestern region, where maize was grown extensively. Maize is a common host for *Fusarium tricinctum* but not for *Claviceps purpurea.*

The region for which we have the most useful health data is Normandy. A Norman physician, Louis Lepecq de la Clôture, published in 1778 his observations of health in his native province in the period 1762–1777. The thousand-page study won him (I think quite justly) a title of nobility.[33] In it are valuable clues to the causes of death and infertility in Normandy, and perhaps in many other parts of France as well.

Despite its relatively advanced agriculture and rich soil, Normandy was not a particularly healthy place. Between 1695 and 1774 the Norman population increased by only 19 percent, as compared with a mean increase of 30 percent for all of France; during 1775–1789 the Norman population declined by 4 percent, as compared with a mean of 3 percent increase for the rest of France.[34] Normandy had a foggy, rainy climate, very favorable for the growth of molds. The local diet included buckwheat, rye, and wheat products, all subject to infection by *Fusarium tricinctum* and related species. Judging from Lepecq's descriptions of diseases, immune deficiency problems were endemic in the entire region and epidemic in 1777, a year with a very wet spring. The less detailed accounts of disease by other physicians in Normandy before and after Lepecq confirm the seriousness of the problem.[35]

Normandy also had a problem of low fertility. Lepecq reported that in Rouen the women native to the town often had irregular menstrual flows, spontaneous abortions, and miscarriages: they tended to become sterile unless they moved elsewhere. Certain lands around Rouen were too sandy for growing wheat, and rye was a staple among common folk. Moreover, Lepecq recognized the presence of gangrenous ergotism in eastern Normandy.

Among the three parts of Normandy there were interesting contrasts. In the west the population grew by 30 percent during 1700–1787, close to the national average of 28 percent. The staple cereals in most parts of the west were wheat and buckwheat. In central Normandy, where the staple was rye, the population grew by only 2 percent during the century. In the east, where both wheat and rye were consumed, the population grew by 11 percent.[36]

One may ask to what extent the situation in Normandy was representative of that in France. There is evidence that *Fusarium* mycotoxicosis was a major national health problem in this period. The most important evidence of its seriousness in France appears in a report on epidemics in France, 1771–1830, by a panel of physicians to the Academy of Medicine.[37] Given the categories into which diseases were sorted at the time it is impossible to determine how many of the deaths were caused by *Fusarium* toxins, but until proven innocent they are implicated in all of the following categories (as diagnosed in 1833): croup, complicated with gangrenous angina (one-fourth of all deaths); membraneous angina (probably diphtheria) and gangrenous angina, simple and complicated (one-fourth of all deaths); and scarlatina, often complicated with serious angina (one ninth of all deaths).

There has been much discussion about the mystery of the "premature" decline in French fertility in the last two decades of the eighteenth century. If Lepecq's observations about infertility and spontaneous abortions in Rouen prove typical of many regions of France, then it may turn out that low fertility was normal, although it varied somewhat with weather. If French peasants were habitually eating ergoty rye bread there is no need to suppose that they were using condoms or practicing coitus interruptus.

The eighteenth-century national pattern changed little during the first quarter of the nineteenth century. In Normandy, dietary change was slow and child mortality remained high: at Quillebeuf, downstream from Rouen (where much rye was eaten), "convulsions" were the cause of death of 237 of the 324 children who died during 1780–1809.[38] In 1830 infant mortality in rye-eating rural areas around Rouen was one in five, the highest in six areas surveyed; the lowest rate of infant mortality (one in eleven) was the wheat-growing area of Yvetôt.[39]

Especially after the dearth year 1818, dietary change in France accelerated. To meet the growing demand for food in the cities, the more fortunate peasants were able to expand wheat and meat production and to eat white wheat bread themselves. The poor turned more and more to potatoes. In Normandy, consumption of potatoes rose steeply from 1825 to 1845 and increased by 1,070 percent from 1825 to 1909. By 1840 the potato had conquered Normandy.[40]

At the beginning of the nineteenth century the Sologne was famous for ergotism epidemics. Infant mortality exceeded 400 per thousand live births (average for late-nineteenth-century Russia). After 1811, however, traditional ways began to change. The introduction of wheat and potatoes into the diet, reducing dependence on rye, and improvements in agricultural techniques are believed to have been responsible for the elimination of ergotism. In a single decade, the 1840s, the death rate fell from over 40 per thousand to 25 per thousand.[41]

As noted above, long-term climatic changes may be as important to population history as human habits are. The first half of the nineteenth century was a time of cool summers in northwestern Europe. In the declining number of rye-dependent English communities this was a favorable trend: only six summers out of fifty (12 percent) were warm enough to favor ergot alkaloid formation. But in the Paris region, 56 percent of all summers were cool enough for alkaloid formation.[42] Surprisingly, then, the higher growth rates in England may have been partly due to a lucky break with the weather.

Variance in the incidence of mold poisoning, caused by climatic changes and dietary changes, does much to explain both fertility and mortality trends in England and France during 1750–1850. The trends include the steady growth of the English population, which demonstrated rising fertility and falling mortality after 1770, and the erratic and more gradual growth of the French population, which showed falling fertility after 1770 and falling mortality after 1795. A decrease in mycotoxicoses may even explain why the great mortality crises ceased to occur when they did.

The food-poisoning argument also helps to explain why there were great differences in the mortality patterns of different regions and even different parishes: that is, the different regions were characterized by different climates and diets.[43] It can explain the seasonality of mortality: T-2 toxin formation peaked in the spring, and deaths from ergotism peaked in the late summer and early fall. It can explain why conceptions tended to be depressed in the late summer and early fall, when the ergot on the newly harvested rye was at maximum toxicity. And it can explain why, after 1750, conceptions were less and less diminished at this time of year in England.

The argument makes clear why it was a decline in child mortality, specifically, that increased population growth in England after 1750: because growing children eat more food per unit of body weight than adults do, they consume more poison per unit weight when the food supply is tainted. Any reduction in the amount of poisoned food tended to save the lives of children. Furthermore, the argument explains class differences in infant and child mortality.

Finally, this hypothesis is generalizable to other parts of Europe and elsewhere. It explains why a population can expand before the introduction of any important new productive technology or medical practices and without any increase in the standard of living of the masses. For all these reasons the food-poisoning argument may be the best explanation for the takeoff in European population growth in the century between 1750 and 1850.

PART III

Contributions to a Health History of Colonial New England

CHAPTER EIGHT

The Throat Distemper

W HEN the British settlers came
to North America they found climatic conditions similar to those of
western Europe (see Figure 19)—with one exception. That was northern New England and points north, where normal weather conditions
resembled those in European Russia. If T-2 toxin were to form any-
where in the New World, one would expect to find it here.

And, indeed, in 1735–1736 in New Hampshire there appeared an
epidemic disease that resembled alimentary toxic aleikiia (ATA),
which is caused by ingestion of T-2 toxin. Known as the throat distemper, this disease was characterized by the necrotic and hemorrhagic symptoms associated with immune deficiency. It was highly
lethal.

In 1939 Ernest Caulfield diagnosed the symptoms as those of
diphtheria and scarlet fever (scarlatina), infectious diseases that were
prevalent at the time.[1] He thought that the disease described in the
country towns of New Hampshire and eastern Massachusetts was
diphtheria and that Boston had suffered an outbreak of scarlatina. But
there were serious discrepancies between the clinical models of
diphtheria and scarlet fever, on the one hand, and the symptoms of
"throat distemper."

First, diphtheria and scarlatina are diseases of the late fall and
winter. The throat distemper was a disease of late spring and early
summer. Second, the distribution of epidemics of throat distemper
was not typical of the spread of contagious disease. For example, in
1736 there was an epidemic of throat distemper in New London, and
none in Groton, just across the river; in 1740 there was an epidemic
in Cambridge, and not in Boston, again just across the river.

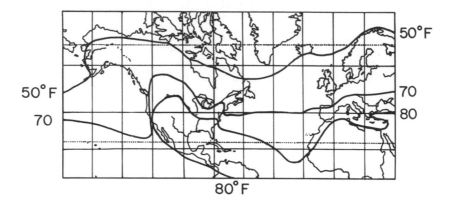

FIGURE 19. Summer temperatures in North America and Europe. Based on maps from Richard O. Cummings, *The American and His Food* (Chicago: University of Chicago Press, 1940), p. 19, by permission of The University of Chicago Press.

Caulfield based his diagnosis on fatality rates. Alleging that the fatality rates in New Hampshire (16–60 percent) were significantly higher than in Boston (2.8 percent), and that the fatality rate of diphtheria was 26–49 percent whereas that of scarlet fever was 3.2 percent, he concluded that the disease in New Hampshire was diphtheria while that in Boston was scarlet fever. In 1939, when Caulfield wrote, scarlet fever was a relatively mild disease. That is why he felt the need to cite diphtheria to account for the high mortality in New Hampshire.

An alternative explanation is available, however: perhaps the population of New Hampshire suffered from weakened immune defenses, in comparison with the people of Boston, and therefore had a higher case mortality rate.

Caulfield used the high case mortality rate among children as an argument in support of the diphtheria diagnosis. Of the 71 fatal cases in Boston attended by a physician who wrote about the epidemic, William Douglass, only nine were over the age of fourteen. But this is not compelling evidence for diphtheria, as Caulfield took it to be, because it is what one would expect in an epidemic of food poisoning as well as scarlatina *or* diphtheria. Furthermore, Caulfield mistakenly attributed to Douglass the report that throat distemper patients in Boston had a skin rash whereas those in country towns did not, but this is what Douglass said:

It [the disease] is generally in so considerable a degree more favourable in Boston, than in the Townships where it first prevailed that many can scarce be persuaded of its being the same Distemper. It is nevertheless essentially the same, *there is no Symptom, even the most malignant that has appeared in New Hampshire, but what the like has occurred in Boston* [italics added].[2]

In short, Caulfield, far from being discriminating, perpetuated the errors of nineteenth-century physicians who, lacking laboratory tests, often reported cases of diphtheria and scarlatina combined or as a single disease. A severely infected throat and scarlet fever might indeed be combined, but the throat condition was often not a symptom of diphtheria. To make a differential diagnosis of diphtheria there must be evidence of a false membrane in the throat. Eyewitnesses of the "throat distemper" *never* reported such a membrane: instead they reported a gangrenous or ulcerous sore throat. The actual explanation may have been that scarlet fever accompanied this throat condition because an immunosuppressant triggered an opportunistic infection by streptococcus A, the bacterium that causes scarlet fever, harbored in the mouth and throat.[3]

Although T-2 toxin has been identified only in overwintered grain in the Soviet Union, in the United States it has been discovered in stored corn. Specifically, it has been found in corn that ripened late the preceding summer, that was especially moist, and that was placed in conventional cribs without artificial drying. The toxin appeared early the following spring.[4] The primary sources concerning the throat distemper epidemic of 1736 indicate that the symptoms fit the model of alimentary toxic aleikiia remarkably well. Corn was an important food source in New England, and if it was tainted with T-2 toxin it could have caused an epidemic like the one described.

In the early stage of the disease many patients complained of a "copperish taste or peppery smart in the throat." The three most important symptoms that characterized the acute phase of the distemper were: an ulcerated and necrotic, painfully sore throat; skin eruptions that were bright crimson, miliary, purpuric, or ulcerous; and a tendency to bleed, surfacing in the form of hemorrhagic areas on the skin, bleeding from the nose, mouth, and uterus, or passing bloody urine and bloody or dark stools. In addition there might be pulmonary and central nervous system complications. Swelling of the oral cavity and neck glands might be so severe as to cause strangulation.

Douglass did not distinguish phases in the course of this disease: its onset appeared to him in 1736 to be sudden, or followed a day or two of mild symptoms. But in 1739 he observed that "the seminium seems to be hatching some time in the blood before the Distemper notoriously discovers it Self."[5] The moribund patient complained of oppression and stricture in the upper chest, breathed with difficulty, and had a deep, hollow, hoarse cough. Strangulation was the cause of about 10 percent of deaths; the other deaths were blamed on "fever," "necrosis," and consequential ills.

In his 1736 account of the throat distemper Douglass denied the existence of relapses, but by 1739, when he wrote a description of the disease to Cadwallader Colden, he had changed his mind and declared that there were some second seizures after a year or two. This would tend to support a diagnosis of poisoning rather than infection. Jonathan Dickinson, reporting on a similar epidemic in New Jersey, said that relapses could occur as often as four times in a year.[6]

Persons who were well-nourished were less susceptible to the throat distemper. Douglass believed that mortality was higher in rural areas because the diet there included coarse food and salt pork, which meant that psoriasis (scaly red skin lesions) was endemic. Given the importance of corn in the New Hampshire diet, this pattern may have been an indication of pellagra. Regina Schoental has suggested that mycotoxins may be a factor in pellagra.[7]

Douglass also blamed high mortality on the presence of lowland "damps" in the vicinity of woods and ponds. This is the sort of environment favorable to fungal growth. Finally Douglass, observing that practitioners attending the sick did not infect themselves or their families, expressed the opinion that the throat distemper was not contagious.

There are a few discrepancies between the symptoms of ATA and that of the throat distemper. Douglass does not divide the course of the disease into a succession of phases, although he indicated the possibility of a long gestation period. If the throat distemper was indeed an instance of ATA, the causal relationship between early symptoms and the acute phase might easily have been missed, for the early and severe symptoms occur about a month apart.

Another discrepancy is the symptom of "ichorous" (thin, serous, or bloody) discharge from the nose, cited in all eighteenth-century accounts of the disease. This discharge caused ulcerations of the skin both of patients and their nurses. This symptom is not mentioned in English summaries of Russian literature on ATA in the twentieth century, but it has been reported that an ether or alcohol extract of the cereal harboring T-2 toxin placed directly on human skin causes inflammation and swelling and that the strongest reaction occurs on the skin of healthy humans.[3] Ulcers on the lips, nose, and fingers in some Russian cases of ATA might have been caused by an ichorous nasal discharge. Because strains of fungi, and the level and kind of metabolites they generate, vary through time and by locality, it would not be strange if there were some variance in the symptoms they produced. Different strains of the same species of fungi might produce an ichorous nasal discharge in an eighteenth-century American epidemic but fail to do so in a twentieth-century Russian epidemic.

The epidemiology of the "throat distemper," as well as its symptoms, fits the model of alimentary toxic aleikiia. The throat distemper first appeared in Kingston, New Hampshire, on May 20, 1735. Reverend Jabez Fitch recorded that in the state of New Hampshire between May 1735 and July 1736 a total of 984 persons died of the disease, of whom 800 were children under ten.[9] The epidemic appeared in Boston in the fall of 1735 and peaked around March 1736. In New London, Connecticut, it appeared in May 1736, peaked in August through October, and lasted until the end of the year.[10] There were outbreaks in Maine in 1737 and 1738,[11] and one in New Jersey in 1735.[12] In Stratham, New Hampshire, an outbreak occurred in 1742.[13]

Weather conditions in New England during the epidemic years may be ascertained from the diary of Joshua Hempstead of New London. The winter of 1734–1735 was mild, with alternating freezing and thawing. The spring of 1735 was wet and chilly, and crops were late in ripening. The winters of 1735–1736 and 1741–1742 were also mild, with alternate freezing and thawing.[14] This was very favorable weather for the production of T-2 toxin.

There is one epidemiological discrepancy, however: most of the epidemics of ATA in Russia were over by the end of July, whereas the "throat distemper" continued through fall and winter. This discrepancy may be explained by the fact that the Soviet authorities were in a better position than the colonial authorities to identify contaminated grain and replace it.

If the symptoms of the throat distemper in New England were caused by T-2 toxin, then how did it form, and what grain did it contaminate? Of the cereals grown in New England, rye was the best substrate for the formation of T-2 toxin. Wheat and white corn tied for second place. Since rye may be the crop of choice in a cold climate, it would have been suited to northern New England (New Hampshire, Vermont, and Maine), where grain was stored in small barns, less than twenty feet square, usually without artificial drying.[15] Under these conditions T-2 toxin will very likely thrive.

If toxin in grain was a necessary cause of throat distemper, then one would expect that the outbreak of 1735–1742 would not be an isolated case. Indeed it was not. An epidemic of throat distemper was reported in the town record of New London, Connecticut, in 1689. The same symptoms, labeled "angina maligna," appeared in Philadelphia in 1746.[16] There were epidemics of the disease in New England in 1754–55 and 1784–1787.[17] As "cynanche maligna" they appeared in the Salem area in 1793.[18] As late as the 1830s the town clerk of Andover, Massachusetts, was attributing deaths to "throat distemper."[19]

The eventual disappearance of the disease may have been a result of dietary changes. Potatoes tended to replace rye as a staple in northern New England. New England, and the nation, came to rely on wheat from the rich lands of the Middle West, where the harvested crop tended to be relatively dry, unfavorable to the growth of molds.

The throat distemper epidemic of 1735–36 in New Hampshire may be a case of acute immune deficiency. Alternatively, a chronic immune-deficient condition in certain areas where the inhabitants ingested sublethal doses of natural immunosuppressants may have contributed to the severity of a variety of infectious diseases.

There is evidence that colonial New England also suffered from ergotism. Ergotism may have been responsible for the condition called "fits" in colonial records. Of particular historic interest are the epidemic "fits" associated with bewitchment and religious "awakening," the topics of the next two chapters. As I have shown for the epidemic of "throat distemper," the possible role of food poisoning must not be overlooked in explaining these events.

CHAPTER NINE

Ergot and the Salem Witchcraft Affair

THE Salem witchcraft affair of 1692 was peculiar. In terms of the number of people accused and executed, it was the worst outbreak of witch persecution in American history, affecting not only Salem Village but eight other communities of Essex County, Massachusetts, as well as Fairfield County, Connecticut. The timing of the outbreak was strange, too—forty-seven years had passed since the last epidemic of witch persecution in England. No one has been able to explain why it occurred in 1692 and not some other year, or why it happened in Essex and Fairfield counties and not in other counties.

The Salem witch trials have been the subject of scores of books, both scholarly and popular, and the problem has been approached from a variety of disciplines. In recent years attention has shifted from the dramatic stories of individuals—the slave Tituba from Barbados, the hysterical adolescent Ann Putnam—to the patterns displayed by the different factions to the dispute. In 1976 psychologist Linnda Caporael proposed that those who displayed symptoms of "bewitchment" in 1692 were actually suffering from ergotism. Caporael matched the symptoms and their epidemiology in 1692 with those in the ergotism model.[1] Psychologists N. K. Spanos and Jack Gottlieb criticized her work on the ground that the facts of the case fit the model very imperfectly.[2] Other evidence, however, supports the position of Caporael and not that of Spanos and Gottlieb.

The historians Paul Boyer and Stephen Nissenbaum, who were concerned with the social reactions to the symptoms of bewitchment rather than the origin of the symptoms, showed that there was a nonrandom distribution of witch accusers. The accusers were mainly supporters of the minister, Reverend Samuel Parris, whereas opponents of the witch trials were overwhelmingly anti-Parris. The accused witches, however, Boyer and Nissenbaum admit, were not associated with the anti-Parris group; they were political outsiders.[3]

In any case one would expect that the distribution of witch accusers and defenders would be nonrandom. Every community has its pecking order, its factions and outsiders. The main problem is to explain the kinds of behavior that were attributed to witchcraft. Boyer and Nissenbaum did not do this.

In *Entertaining Salem*, John Demos discussed not the Salem affair specifically, but the witch cases that appeared sporadically in other years. He showed that the distribution of persons accused of witchcraft was nonrandom. This we would also expect. He devoted several pages to the physical illness and animal symptoms attributed to witchcraft and reveals that out of fourteen victims fatally afflicted, nine were under ten years old. He made no effort to explain this nonrandom age distribution, however, and he did not discuss the ergot theory.[4]

The suggestion that the afflicted teenage girls in Salem Village were feigning their symptoms or, as Spanos and Gottlieb suggested, role playing in the presence of social cues, cannot explain the symptoms of the animal victims who were apparently not stimulated by social cues. Nor can feigning or role playing explain the deaths that occurred among the "bewitched."

The victims of bewitchment did not have true convulsions, as Spanos and Gottlieb pointed out, because in a strict sense today convulsions imply loss of consciousness, but this does not disprove a diagnosis of ergotism. Victims of dystonic ergotism writhe and have spasms but normally do not lose consciousness. The adjective "convulsive" used to describe this form of ergotism has been used loosely, in the sense of "spastic."

It has also been suggested that the bewitched were suffering from hysteria.[5] This explanation, too, is unsatisfactory. People in the afflicted communities may have been hysterical in the sense that they were excited and anxious, but psychological stimuli alone have not been shown capable of producing an epidemic of severe central nervous system disturbances, deaths, and animal damage.[6] Yet another proposal, that the symptoms of bewitchment were those of Huntington's chorea, is not supported by the evidence of the age of the victims. That disease does not become manifest until middle age, whereas the bewitched of Salem were mainly young.[7]

What did New Englanders label as "bewitchment"? In Essex County, Massachusetts, twenty-four of thirty victims of "bewitchment" in 1692 suffered from "fits" and the sensations of being pinched, pricked, or bitten, all of which are common symptoms of ergotism. According to English folk tradition, these were the most common specific signs of bewitchment.[8] These were the symptoms most often mentioned in the court records, for the intent of the court proceedings was to prove witchcraft, not to present a thorough medical case history.

Some of the other symptoms of bewitchment mentioned in the court record are characteristic of ergotism. These include temporary blindness, deafness, and speechlessness; burning sensations; visions like a "ball of fire" or a "multitude in white glittering robes"; and the sensation of flying through the air "out of body." Three girls said they felt as if they were being torn to pieces and all their bones were being pulled out of joint. Some victims reported feeling "sick to the stomach" or "weak," sensing a "burning" in the fingers, swelling and pain in half of the right hand and part of the face, and being "lame."

It should not be surprising that gastrointestinal symptoms were rarely mentioned. Reformatskii, in his study of dystonic ergotism in northeastern Russia (1889–1890), found that gastrointestinal symptoms were frequently absent.[9] The Salem court record does register some gastrointestinal complaints, but such complaints were nonspecific to the legal condition known as bewitchment and therefore of little interest in supporting a charge of witchcraft. One would not expect them to be prominent in court records

According to the Salem court record three people and several cows died. Psychological suggestion cannot explain such deaths, and there is no reason to exclude these cases from the damage attributed to witches. They cannot be dismissed as ordinary occurrences stemming from any of several possible causes. Any event taken in isolation may have many different possible causes: it is the pattern that is important.

What follows are some significant passages from the Salem court record. The first is an account of human sickness and cattle death on one Essex County farmstead. According to the testimony of Sarah Abbott, August 3, 1692:

> My husband, Benjamin Abbott has not been only afflicted in his body, as he testifies, but alsoe that strange and unusual thing have happened to his Cattle, for some have died suddenly and

strangely, which we could not tell any natural reason for . . . and some of the Cattle would Come out of the woods w'th their tounges hanging out of their mouths in a strange and affrighting manner . . .[10]

Nathaniell and Hannah Ingersoll described the case of a death in which no witchcraft was suspected:

We ware conversant with Benjamin Holton for about a week before he died and he was acted in a very strange manner with most violent fitts acting much like to our poor bewitched parsons w'n we thought that they would have died tho' then we hade no suspition of witchcraft. amongst us and he died a most violent death with dreadfull fitts and the Doctor that was with him said he could not tell what the distemper was and he died about Two days before Rebekah Sheepard [who died after suffering similar symptoms].[11]

In the John Putnam household both a parent and child were ill with the same symptoms; the child died:

The Deposition of John Putnam weaver: and Hannah his wife who testifieth and saith that our child which dyed about the middle of April 1692; was as well and as thriving a child as most was: tell it was about eight weeks old . . . Mary Estick and Sarah Cloyes and I myself was taken with strange kinds of fitts: but it pleased Allmighty God to Deliver me from them: but quickly after this our poor yong child was taken about midnight with strange and violent fitts . . . it continewed in strange and violent fitts for about Two day and Two nights and then departed this life by a cruell and violent death being enuf to piers a stony hart. for to the best of our understanding it was near five hours a dying.[12]

Social cues in the courtroom may have stimulated some of the hallucinations reported, but outside stimulation does not disprove a diagnosis of ergotism. Ergot is the source of lysergic acid diethylamide (LSD), and it may include natural alkaloids that act like LSD. People under the influence of this compound tend to be highly suggestible. They may see formed images—for instance, of people, animals, or religious scenes—whether their eyes are open or closed.[13] These hallucinations can take place in the presence or absence of social cues.

Symptoms similar to those mentioned in the Salem court record also appeared between May and September of 1692 in Fairfield County, Connecticut. A seventeen-year-old girl, Catherine Branch, suffered from fits, pinching and prickling sensations, hallucinations, and spells of laughing and crying. On October 28 she died, after accusing two women of bewitching her. John Barlow, age twenty-four, reported that he could not speak or sit up and that daylight seemed to prevail even at night. He had pain in his feet and legs.[14]

The victims of bewitchment in Essex County were mainly children and teenagers. Seven infants or young children are known to have developed symptoms or to have died. According to recent findings, nursing infants can develop ergotism from drinking their mother's milk if she has the disease. It has also been found that adolescent victims of ergotism were especially susceptible to mental disturbances.[15]

Spanos and Gottlieb, citing the court record, asserted that the proportion of children among the victims in 1692 was less than that in a typical ergotism epidemic. In a recent epidemic of gangrenous ergotism in Ethiopia, however, the ages of the victims were not much different from those in the Essex County epidemic of 1692: more than 80 percent of the Ethiopian victims were between the ages of five and thirty-four.[16] The difference in symptoms in the two types of ergotism is primarily a function of which alkaloids are produced by a given *Claviceps* strain—not the age of those exposed to ergot toxins. The age incidence of ergotism depends on how much rye the victims eat and how it was prepared. Rigidly rye that is boiled rather than baked is less toxic and may be the preferred form given children. Physically active teenagers may consume more food per unit of body weight than young children do. Factors such as these may, singly or in combination, shift the age distribution of the disease by a significant number of years.

There can be no doubt that rye was cultivated in Salem Village and in many other parts of Essex County in the late seventeenth century.[17] The animal cases could have resulted from ingestion of wild grasses such as wild rye or cord grass, found in Essex County and vulnerable to ergot infection.[18] It was recently discovered that ergot alkaloids are ubiquitous in tall fescue pastures in Georgia, and that ergotism-like toxicoses may be caused by fungi other than *Claviceps purpurea*.[19]

Was there ergot in the rye of Essex County in 1691–1692? That is a critical issue. As explained in Chapter 1, rye bread made from bolted (white) flour with an ergot content of 3 percent or more is cherry red in color. Three Essex County women who attended a witch sacrament in the meadow of Reverend Parris (Salem Village) declared that the sacramental bread was red.[20] This may be interpreted in three ways. (1) It was a random occurrence of no significance. Then why was it mentioned at all? (2) Red rye flour was common in the village at that time without anyone intending it to be so. (3) A red pigment was added deliberately to the bread intended for the witch sacrament. If this red pigment was ergot, it might serve to bring on hallucinations during the sacrament. Such evidence of the presence of ergot in itself is not conclusive, but given the epidemiological evidence it is interesting at the very least.

Epidemiologically, the timing of the appearance of the symptoms was important. The first symptoms of bewitchment appeared in Salem Village in December 1691. Beginning about April 18, 1692, the pace of accusation increased. It slowed in June and then reached a peak between July and September. Exactly when the symptoms terminated is unknown. After October 12, 1692, there were no more trials for witchcraft by order of the governor of Massachusetts. During the winter of 1692–1693, however, there were religious revivals in the area around Boston and Salem, during which people saw visions.[21]

Many questions arise concerning the timing of these events. If rye harvested in the summer of 1691 was responsible for the epidemic, why did no one exhibit any symptoms before December of that year? In the ergotism epidemics of continental Europe, the first symptoms usually appeared in August, immediately after the rye harvest. But European epidemics occurred in communities heavily dependent on rye as a staple crop and among people so poor that they had to begin eating the new rye crop immediately after the harvest. The situation was otherwise in New England. Recent research indicates that it was normal to delay consumption of new crops until fall.[22] Probate inventories and the diary of Zaccheus Collins, a resident of the Salem area during the epidemic, show that if other food was abundant, the rye crop often lay unthreshed in the barns until November or December.[23] Since ergot can remain chemically stable in storage for up to eighteen months, stored rye might have been responsible for the symptoms of December 1691.

But if people normally delayed threshing rye until winter, why was there a peak in disorders of the central nervous system in the summer of 1692? Was food scarce at that time—so scarce that the new rye crop was immediately utilized?

Unfortunately, the usual source of information about food dearth, Massachusetts government records, is missing for 1692 for political reasons. One may, however, make use of contemporary diaries. Lawrence Hammond wrote that on March 1, 1692.

> These rains and ye Violent sudden melting of ye snow in ye Wildernes caused such a sudden & Violent flood yt hath done abundance of damage in most parts of ye Country, carrying away bridges, Mills, etc. Connecticut river 3 ft. higher yn ever it was known before, destroyed mch Cattle in ye meadows, carryed away some Houses, & washed away in many places ye very land with ye English graine sown in it.[24]

So the first crop damage came in early spring. Later there was drought. The Massachusetts General Court ordered a public fast on May 26, 1692. Samuel Sewall wrote that on July 27 there was "plentiful rain after a great drought," and on August 25 a fast was ordered against both drought and witchcraft.[25]

There was also evidence of plant disease. To the south, the Colonial Assembly of New York ordered fasts monthly from September 1691 to June 1692 because of "a burdensome war and a blast upon the corn" ("corn" meant grain in general; "Indian corn" meant maize, what we know as corn).[26] Such evidence alone does not prove that there was a dearth of food in Essex County, Massachusetts, in the summer of 1692, but it does establish extensive crop damage.

More evidence comes from data about climate, as indicated by tree growth.[27] Tree-ring measurements in Nancy Brook, New Hampshire, tell us a great deal about weather in eastern Massachusetts. The thickness of the rings reflect variation in temperature, for the amount of moisture in the area even in dry years was usually sufficient for tree growth. In New England, the optimum temperature for ergot alkaloid production (17.4°–18.9° C) would be reached only in an unusually cold growing season. Very narrow rings may indicate such extreme temperatures.

Table 10 shows that in the period 1660–1689, the year with the coldest growing season (and therefore the lowest tree-ring index) was 1688, but 1688 did not have the cold preceding winter needed for ergot formation, and the summer may even have been *too* cold. The rings for 1690 and 1691 indicate unusual cold and for 1692, below-normal temperatures. The years 1693–1697 were also cold, but because witchcraft prosecution was prohibited in Massachusetts after 1692 we cannot use these years to test hypotheses.

TABLE 10. Tree-ring indices in Nancy Brook, New Hampshire, 1660–1759

Decade	Year									
	0	1	2	3	4	5	6	7	8	9
1660	80	76	96	68	126	85	108	138	111	161
1670	100	123	102	118	114	105	133	108	115	109
1680	101	100	85	83	87	113	107	83	57	81
1690	57	62	74	64	64	67	71	61	75	89
1700	99	104	171	136	121	136	122	163	167	117
1710	95	95	134	168	164	141	143	132	109	115
1720	81	93	109	97	109	122	107	102	67	99
1730	93	110	111	87	118	100	89	82	90	90
1740	83	65	69	74	78	86	93	99	76	83
1750	80	72	83	91	101	93	98	86	94	101

Source: E. DeWitt and M. Ames, eds. *Tree-ring chronologies of eastern North America* (Tucson, 1978), p. 19.

If rye was infected by *Claviceps purpurea* in 1691 and 1692 in New England, why would it not be infected in the middle and southern colonies? Why was witchcraft persecution limited to New England? The most likely answer is that, because New England had colder winters and cooler summers than its neighbors to the south, it was more susceptible to ergot poisoning, whose symptoms were taken to be a sign of bewitchment. Summer temperatures favorable for ergot alkaloid production occurred in Cambridge, Massachusetts (where the July mean temperature was 18.2° C, 1743–1774), but even as far south as New York City (July mean, 1789–1815, 23.2° C) the summers were too warm (see Figure 19).

The spatial distribution of symptoms within Salem Village is also relevant to the case for an ergotism diagnosis. The growth of population in the area provided an incentive for farmers to utilize swampy, sandy marginal land. This land, if drained, was better suited to the cultivation of rye than other cereal crops. But low-lying, wet land was land on which rye was most likely to be infected with ergot, and rye grown on newly cultivated land is also more likely to be infected.[28]

All twenty-two of the Salem households with symptoms in 1692 were located on or at the edge of soils ideally suited to rye cultivation: moist, acid, sandy loams (see Figure 20). Of these households, sixteen were close to river banks or swamps and fifteen were in areas shaded by adjacent hills, and therefore moist. No part of Essex County is more than 120 meters above sea level. As in Essex County, the pre-

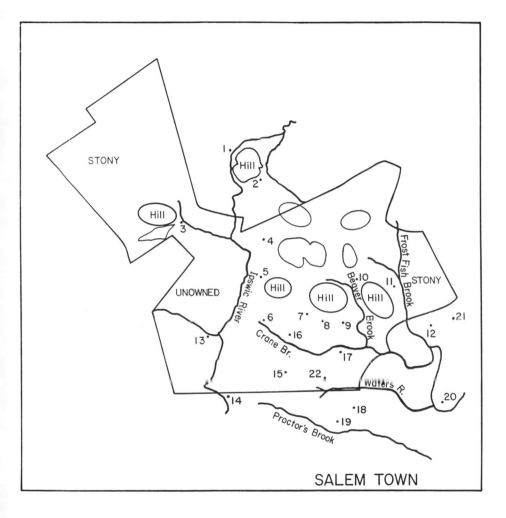

FIGURE 20. Location of dwellings of the bewitched in Salem Village.

dominant soil type in southern Fairfield County, Connecticut, was fine sandy loam, and elevations were low.[29]

It must be noted that rye was not the crop of choice of the settlers. Beginning in the 1590s, the common people of southeastern England began to eat less rye bread. When the settlers first arrived in New England they planted wheat, but the crops were troubled by wheat rust and in the 1660s the farmers began to substitute rye for wheat.[30] During 1664–1689 they were lucky: weather conditions did not favor *Claviceps* infection of the rye, but in 1690 a cold trend began and the latent danger of ergot infection materialized.

The physical, mental, and social disturbances that plagued Salem Village in 1692 were not unique to that time and place. Nor would we expect them to be if they are, in fact, symptoms of a not uncommon disease. In 1653 in New Haven, Connecticut, for example, a woman fell into "strange fits" and had a ravenous appetite. Another woman had fits, felt she was pinched, heard a hideous noise, "and was in a strang manner sweating and burning and some time cold and full of paine yt [that] she shriked out."[31] In December 1680 in Newbury, Massachusetts, the victim suffered from fits, feeling pinched, fainting, nausea, and a ravenous appetite.[32] The combination of strange sensations with a ravenous appetite is common in ergotism. After 1692 rye consumption increased and there was cold weather aplenty, but there were no more witch trials in Massachusetts—the authorities forbade them. The disturbing symptoms continued to occur, but another interpretation, not wholly new, was culturally mandated. This will be the subject of Chapter 10.

Although the limitations of surviving records make certainty impossible, the balance of the available evidence suggests that witchcraft accusations in 1692 in New England were prompted by an epidemic of ergotism. The Salem witchcraft affair may have been a reflection of a largely unrecognized but endemic health problem in the New World.

CHAPTER TEN

Great Awakening or Great Sickening?

During the Great Awakening, a religious revival that reached a peak in New England in the fall of 1741, hundreds, perhaps thousands of people experienced fits, trances, and visions. Although individuals and even whole communities had experienced such symptoms before, in 1741 the number involved was greater than at any time previous. No satisfactory explanation has been proposed.[1]

I do not intend in this chapter to interpret the Great Awakening as a whole, nor to explain why central nervous system disturbances were regarded as divinely inspired. By drawing attention to the unusual symptoms and their epidemiology, I hope to explain only why the Awakening occurred when and where it did. There are two questions at issue here, and they need to be kept separate: what caused the symptoms, the focus of concern in this chapter, and what caused the interpretation given those symptoms? The approach is analogous to focusing attention on the harm allegedly done by witches (fits, hallucinations, and so on) and keeping this part of the problem distinct from the social response to the actual harm done. My comments apply only to revivals in which abnormal symptoms of the central nervous system appeared: namely, in New England in the eighteenth century and on the Kentucky frontier at the beginning of the nineteenth century. There were religious revivals elsewhere that were characterized by excitement but not spasms and hallucinations. My theory has nothing to contribute to the explanation of such revivals.

A clue to the diagnosis of the symptoms of the Great Awakening of 1741 may be the fact that in the summer and fall of that year certain parts of New England were suffering from what was perceived as an epidemic disease: nervous fever, as it was called then.[2] Is it possible that this disease, and the physical and nervous symptoms associated with the Great Awakening, may have had one and the same cause—namely, ergot poisoning?

A parallel event occurred in northeastern Russia. N. N. Reformatskii, in a study of the mental symptoms associated with an ergotism epidemic in Viatka province in 1889–1890, found that 39.1 percent of the hospitalized sick suffered emotional traumas, and he believed that the proportion in the population at large was higher. The most common mental symptoms were trembling, confusion, fluctuating moods, attacks of excitement persisting for days and even weeks, and both unpleasant and pleasant hallucinations. Victims of unpleasant hallucinations may have believed themselves drowning or may have seen a fire and feared being burned alive; some thought they saw the Devil himself, demanding that they give up their faith. Those who had pleasant hallucinations, on the other hand, saw, in the persons around them, gods, saints, and members of the nobility; in consequence, the victims were overcome with joy, prayed to these figures, sank down to their knees before them, and believed they were in paradise. After recovering their senses—for between episodes there were remissions—they remembered these pleasant experiences and believed they had really been in paradise.[3]

Similar experiences are found among people who ingest LSD, an ergot derivative, some of whom report that afterward their religious beliefs grew stronger. After recovering from an experimental LSD experience, forty-two persons were questioned: of these, 60 percent stated their religious beliefs had changed; 60 percent said they trusted God more; 30 percent said they felt greater conviction about the existence of a Supreme Being. Out of a group of over 100 subjects in Baltimore, 75 percent said they had intense mystical experiences during their first session with LSD.[4] Such interpretations of LSD experiences were, of course, culturally conditioned both by religious traditions and by the environment of the 1960s.

Many mycologists believe that LSD can appear in ergot naturally, without artificial processing,[5] but this has not been demonstrated. Nevertheless, other potent hallucinogens—ergine, ergonovine, and lysergic acid hydroxyethylamide—are commonly found in natural ergot.[6] We must ask, then, if the disturbances spreading throughout New England in the middle of the eighteenth century match the effects of these hallucinogens found in ergot.

THE SYMPTOMS

New England physicians of this period did not, apparently, keep case records. Nor are there any statistics of morbidity. Various diarists, however, both clerical and lay, provided artless descriptions of the symptoms experienced by themselves, members of their families, and their neighbors. While no single source is sufficient, the combined sources provide enough information for a diagnosis to be attempted. Church records of reported conversions provide a basis for estimating morbidity. In addition, there is useful information about the epidemiology of the symptoms.

The repertoire of symptoms reported in 1741 and 1742 included muscular contractions and spasms; fainting fits followed by a stupor ("lying as if dead") that may have lasted for many hours; hallucinations, such as being "out of body," seeing "a great light in the night," visiting Heaven and Hell; sensations of burning heat and terrible cold; trembling and twitching; numbness; difficulty in speaking and speechlessness; weakness; generalized pain; uterine contractions; leg pain, lesions, and lameness; slow and painful discharge of urine; nausea, vomiting, and diarrhea; ravenous appetite during remissions; moods of joy, despair, emptiness; and various violent and demented behavior.

According to Reverend Charles Chauncy, an opponent of the revival movement:

> When they come out of their Trances they commonly tell a senseless Story of Heaven and Hell, and whom and what they saw there ... In some towns, several Persons, both Men and Women, that formerly were sober, and to all Appearance truly pious, are raving distracted, so that they are confined and chained. Many fall into Epilepsies, as they walk in the Streets, or in their Houses ...
>
> [Many people experienced] a sudden and terrible Fear of divine Wrath, or the Miseries of Hell, occasioning a Sensation of Cold, in most a very extraordinary warmth all over the Body; causing People to cry as if distracted; to shed Tears in great Plenty; throwing many into Convulsions, and a few for some Time into Despair.[7]

Not every person afflicted in 1741 had all the symptoms mentioned, as is true of many epidemic diseases. This is especially true in epidemics of ergotism, in which symptoms often vary a great deal from one patient to another.

The vasoconstrictive property of ergotamine may cause lameness or necrosis (gangrene) in the extremities. This symptom, together with nervous symptoms, did indeed appear in 1741. The Reverend James Davenport, the most emotional of the revivalist preachers, stated in 1744 that at the time of his extravagant behavior his body, especially his leg, was much "disordered" with a "cankry humor" (swelling and ulceration).[8] Joshua Hempstead reported in his diary (August 21, 1741), "I am Something Lame with ye humors falling down into my Leggs but small sores."[9] Reverend Stephen Williams reported in September 1741 that both he and his wife were lame.[10] Reverend Marston Cabot of Thompson, Connecticut, also recorded he was lame.[11]

In one case, that of the Reverend Peter Thacher, central nervous system disorders were combined with "strangury," a slow and painful discharge of urine produced by spasmodic contraction of the urethra and bladder. Renal spasm, too, may be a symptom of ergotism.[12]

Behavioral changes were also noted. Charles Chauncy collected cases of violent and rebellious behavior: attempted suicide and murder; insubordination of children against parents, servants against employers, and laymen against clergy.

Most published reports of symptoms during 1741–1742 described either psychological symptoms or physical symptoms, but not the two together. The following account, a letter of Henry Lyon, from Lebanon, Connecticut, dated March 12, 1742, presents both in a single concept:

> on Tuesday Last I being at a meeting mr. Pumry [Pomeroy] preached from the 40 chapter of Isaiah—1 and 2 verses I found my hart in sum measure drawn forth to God sermon being over my soul whas filled with ravishing transport to the doore it seemed then whas nothing but a thin paper wall that separated me from perfect Glory my visions increased and my hart still reaching forth after God at last I fainted . . . [many visions follow] . . . then my senses returned to me and I found my body all disordered with the cramp.

This letter is annotated in a more literate hand: "drenched in sweat & Limbs drawn up with ye Cramp."[13]

In the diary of Reverend Stephen Williams of Longmeadow, Massachusetts, we find:

JULY 7. In ye aftern[oo]n I went to Suff[iel]d—where I heard
of ye remarkable outpouring of ye Spirit of G.d ... [that] on ye
Sabboth there were 95 P[er]sons add[ed] to ye ch[urc]h

Abt Sun—an hour higher—we had an Exercise—on ye meeting House-Hill ... ye congregation—remarkably attentive &
Grave— ... in ye Evening ... there was considerable crying
among ye people—in one part of ye House or another—yea—&
a Screaching in ye streets—one woman—come to ye House
where—I lodg[e]d ... [that] was Greatly destressd—but She
Gave—a very impfect acct of things ...

JULY 8 ... then went over to Enf[iel]d—where we met Dear
Mr. E. of N.H.—who preached a most awakening Sermon from
those words Deut—32.35—and before ye Sermon—was done
there was a great moaning—& crying out throughout ye whole
House—what shall I do to be Savd—oh I am going to Hell—oh
what shall I do for a christ ... So ... [that] ye minister—was
obligd to desist—shreiks & crys—were piercing and Amazing ...

JULY 9 [Longmeadow] there were considrably shakeing &
trembling before ye revivle was finished ... Sat in ye Allay before
ye pulpitt—[that] was all ... twas dark when we came out of ye
House ... but presently I heard—a crying out in ye yard—when
lo—I found my own Son John Speaking freely—Boldly—& Ear-
nestly to ye people—& warming of them agst Damnation—&
inviting them to [Christ]. I Spoke to him—but—he Seemd beyond himself—& had great discoveries of ye love of [Christ] &
had a great concern—for Souls—he [said] he wondered at himself
... John—Seemd weak & faint—but he came Home with us—
and we had a comfutable night.

In some localities an epidemic sickness of unknown nature broke
out. Williams wrote in his diary on August 12, 1741, "We kept as a
Fast in ye parish upon ye acct of ye sickness," and on October 12,
"this day I hear of Deaths and Sickness in Severall places." Reverend
Daniel Wadsworth of Hartford referred on September 24 to "the dis-
tressing sickness among us." Reverend Elisha Williams of Wethers-
field wrote on October 10 to his half-brother, Reverend Solomon
Williams, "It is not easy to conceive ye Distresses when so many are
sick at once & needing help impossible to be obtained." Shortly before this time Eunice, Elisha's twenty-five-year-old daughter, had
died. "Histerick disorders," he said, "robbed her understanding of its
just exercise some Hours before her death." At the same time his

child "Bille" was suffering from "fever and flux."[14] What is significant is that two children in the same family were seriously ill with symptoms characteristic of ergotism.

The diary of Joshua Hempstead of New London tells the precise timing of the onset of symptoms in 1741, the course of events, and the environmental context.[15]

> JUNE 5. Since I have been absent on the first day of June, John Minor of Lyme Died of a Consumption & was buried ye 2nd also there hath been the wonderfull work of God made Evident in the powerfull Convictions Conversion of Diverse persons in an extraordinary manner which began at the preaching of Mr. Mills of Derby the last Sunday in the afternoon & Several Lectures by him & mr. Eells of Stonington, that this week hath been kept as a Sabbath most of it, & with the greatest Success Imaginable & beyond what is Rational to Conceive of it by the Account that I have of it by all hands. Never any Such time here & Scarce anywhere Else . . .
>
> JUNE 25. mallacholly news of the . . . of Abiah Southmayd formerly Douglass the Eldest Daughter to Richard Douglass Decd. She left Six children had been troubled . . . about 2 months and put an End to her trouble & threw her Self into . . . which was a near neighbors 12 foot water in it and was there drowned . . . night.
>
> JUNE 29. Ruth Tilly wife of James Tilly aged 40 Died at night of a flux vomiting . . .
>
> Mr. allin a young man of New Haven west side . . . Evening Sermon being night first Diverse women being Ter[ri]fied out exceedingly. Dr. Davenport Dismist the Congregation & Some went . . . into the broad ally which was much crowded & there he Screamed out Come to Christ & was his tone & Come away Come away. he got into ye 3d pew on the women's Side & there he held it. Sometime Singing & Sometime praying he & his . . . avers & Mr. allin all took their turns & the women fainting. Mr. Huston . . . ern til 10 Clock att night or thereabouts & then he went off Singing to . . .
>
> I went to Stonington to get my Rye Reaped . . . Stephen Bennet Reaped. . . . Son Minors Josh is not well ague & fever . . .
>
> JULY 21. I opened my Rye that was Stackt up 8 Stacks it being too Gr[een] . . .

JULY 23. Mr. Davenport pr att Groton 4 or 5 days & mighty works followed. near 1000 hearers (near ye old meetinghouse) from all Quarters held ye meeting till 2 Clock at night & Some Stayed all night under the oak tree & in the meeting house. about 60 wounded, many Strong men as well as others . . .

Hempstead himself was in good health until August 21, when he first reported that he was lame. The condition lasted through early October. It should be noted that his own rye crop, harvested July 21, was neither threshed nor consumed immediately because it was green and needed to dry.

Nathan Cole of Kensington in Hartford County was a simple man who kept a journal of his religious experiences, dated only by the year. His entries for 1741 record experiences both mental and physical:

There was then a very Mortal disease in the land, the fever and bloody flux; and I was possest with a notion that if I had it I should die and goe right to hell, but I presently had it and very hard too . . .

And while these thoughts were in my mind God unto me and made me Skringe; before whose face the heavens and the earth fled away; and I was Shrinked into nothing; I knew not whether I was in the body or out. I seemed to hang in open Air before God . . . Now while my Soul was viewing God, my fleshly part was working imaginations and saw many things which I will omitt to tell at this time . . .

And at the end of 52 days I went out abroad . . .[16]

In Lebanon, Connecticut, Reverend Jacob Eliot first observed groaning, crying out, and fits in his congregation on July 14, 1741. In December he visited a twelve-year-old boy and girl in a trance, hallucinating. In the middle of 1742 Eliot noted the following episodes under the heading, "Remarkables in Time of new Light":

A story of a Man at the Westward (I think at Brandford or Norwalk) that saw a great Light in the Night had a Strang impression on his mind that he was to glorify God by killing his Wife & 2 of his children & one of his Brothers or sisters—of his attempting to kill his Wife in ye morning of his arm & joints stiffening upon it & of his running distracted afterward—of a Man at Wallingford running into Despair & drowning himself;

of a Woman (Ensign Buel's Daughter) running distracted—of another at ye Crank long remaining so . . .
also of H. Bliss praying so once or twice and reproaching me for
not praying with his children in a Trance when he himself hin-
dered me; of his wishing ye plague might come & spread in ye
Country because ye children of God were safe & it might be
a means to convert yet many—of his not sowing or planting
because he expect ye end of ye world before harvest—ye Spirit
of God in his breast a 1000 times more to him than ye Scripture
Of his feeling Christ in him & knowing ye Spirit of Christ in
him.[17]

Over many years relapses of symptoms, such as weakness, dizziness, tremors, general bodily pain, headaches, depressions, epilep-
tic-like fits, and psychotic episodes were reported.[18] In Durham, New
Hampshire, some Congregational parishioners in 1746 were given to
choreas—compulsive dancing and other bodily agitations.[19]

Could the symptoms have been caused by an epidemic infection
of the central nervous system? A number of viral infections may
produce convulsions and delirium, but such infections do not produce external swelling, ulceration, and necrosis. Epidemic cerebros-
pinal meningitis has similar symptoms, but it is a late winter disease.
The frequent relapses of the victims do not suggest an infectious
origin of the symptoms but point, rather, to a toxin.

Could the behaviors exhibited in 1741 have been produced by
psychological suggestion, such as the emotional preaching of the revivalist ministers? Some symptoms probably were produced in that
way—for example, the crying observed in some subjects. The content
of the visions of many persons was evidently influenced by their
religious beliefs and perhaps also by the preaching; so also were the
interpretations given to such visions. Considering the widespread
incidence of the symptoms in 1741, some by coincidence would be
expected to occur during and after revival meetings; and since people
who were already feeling somewhat indisposed might attend such
meetings in hope of improvement, the occurrence of still more severe
symptoms during the meetings was not surprising.

In some cases the emotional sermons of the preachers may have
been themselves symptoms of disease. We know that such leading
preachers as James Davenport, Nicholas Gilman, Peter Thacher, and
Eleazer Wheelock suffered in the fall of 1741 from various symptoms
noted above. Psychological suggestion simply does not explain enough
of the pattern of excitement in the Great Awakening.

EPIDEMIOLOGY

It is generally agreed that in 1741 rye was a widespread staple crop in colonial New England. As noted in Chapter 9, rye cultivation increased in the 1660s, when the wheat "blast," probably caused by the fungus *Puccinia graminis*, ruined many wheat crops.

By the first quarter of the eighteenth century in Middlesex County, Massachusetts, wheat products were luxuries for important social and sacramental occasions; rye and maize (Indian corn) were the staple starches. In the inventories of wills in which stocks of rye and maize were separately noted, during 1735–1747 rye constituted about two-fifths of the grain inventoried. Potatoes were an insignificant part of the diet.[20]

Some diarists before and immediately after 1741 recorded the fact that rye was being grown on their own farm or in their community. In addition, they sometimes noted the time of the rye harvest and certain outbreaks of illness that occurred at that time. Some examples follow.

Stephen Williams recorded on August 20, 1736, the prevalence of an "awefull Sickness"; mortality was at a peak in October. There were three deaths in December and a case of "fits" on December 29. In the same month the *New England Weekly Journal* reported that in Hampshire County, Massachusetts, "a great many Cattle have lately died there in their Fields, occasioned (as is supposed) by their Eating smutty Corn, of which, they had this Year, a greater plenty than usual" (December 8, 1736). In Lebanon, Connecticut, the Reverend Jacob Eliot noted that the rye had been harvested on August 4, 1737. This was one of the places where the symptoms of "awakening" were widespread in 1741.

Seventeen forty-one was a bad year in terms of food supply, mainly because of damage done to crops and livestock during the severe winter preceding. The wholesale price index for the British colonies was 73.6 in 1741, as compared with 59.6 in 1739–1740, 69.7 in 1742, and 59.7 in 1743.[21] In Salem, Massachusetts, the price of wheat rose from 12 shillings a bushel in 1740 to 21 shillings, in 1741; the price of rye rose from 8 shillings in 1740 to 15 shillings 6 pence in 1741.[22] Stephen Williams of Longmeadow, Massachusetts, and David Hall of Sutton, Massachusetts, made note of famine conditions in their diaries in March 1741.[23] Food shortages occurred in the new settlements along the Housatonic River in western Connecticut[24] as well as in eastern New Hampshire.[25] In Maine the Reverend Thomas

Smith noted on January 10, 1741, "There has been for some time a melancholy scarcity of corn." On May 25: "Corn is rotten in the ground everywhere and a pretty deal that was planted is not fit for seed."[26] In May 1741 the General Assembly of Connecticut forbade the export of grain, flour, and bread from the colony.[27]

The weather in 1741 was clearly favorable to ergot infection. The winter of 1740–1741 was extremely cold—the worst within memory of New England. Thirty snowfalls occurred, as compared with twenty-six in 1697–1698, the worst winter previous. Long Island Sound was frozen solid over three leagues across, so that people from the mainland could ride across every day. As late as April 1 the ice on the Connecticut River was so hard it could be crossed on horseback. On July 9 people were making punch with ice from a chunk, as big as two carts could draw, floating at the tail of a sawmill in the Connecticut River.[28]

Tree-ring data from Nancy Brook, New Hampshire, shows the growing season of 1741 to have been usually cold (see Table 10). There, and in northern New England generally, it may have been too cold during the growing season for ergot alkaloid formation. Only in Durham, New Hampshire, on the coast, where the ocean moderated temperature, did any strange symptoms appear.[29] Symptoms were concentrated, that is, in southern New England.

In 1741, the onset of the symptoms was what one would expect in an ergotism epidemic. Normally rye ripened in July. In June 1741, when Reverend Jonathan Edwards gave his famous sermon, "Sinners in the Hands of an Angry God," at Northampton, Massachusetts, the audience was indifferent. But on July 8, when he gave the same sermon in nearby Enfield, the audience was terrified. Another sermon, preached on July 8 by a minister across the Connecticut River in Suffield, caused as much excitement as Edwards had. Incidence of symptoms peaked during August and September 1741, just after the rye harvest. The symptoms appeared with less frequency during 1742 and the last episode of unusual group behavior, in New London, Connecticut, occurred in March 1743.[31]

The duration of symptoms is understandable in view of the fact that epidemic symptoms of ergotism frequently continue for many months; for ergot toxins, if stored in a cool dry place at 21°–27° C, may deteriorate only negligibly for seventeen months.

Ergot was more likely to develop on rye growing in a low-lying field or valley with sandy, cold, marshy soils. In Connecticut in 1741 the greatest prevalence of symptoms was in towns located at elevations of less than 400 feet (see Figure 21 and Table 11). Woodbury (elevation 264 feet) was affected, but nearby Torrington (595 feet) and Bethlehem (861 feet) were not. The northern part of Windsor County, which was more elevated than the southern part, was less affected, and Woodstock (592 feet) was not affected at all.

Not all lowland areas were equally affected, however. Rhode Island specialized in dairy farming, and here the symptoms of "awakening" were rare. Moreover, old seaboard settlements, such as Providence and Boston, which had no more marginal land to bring under cultivation for their growing population, were less affected; for rye grown on land long and continuously cultivated was less likely to be infected by ergot.[33]

All sources agree that ergotism is more likely to occur in children and teenagers than adults. Although no age group was exempt, the typical person exhibiting the symptoms was a child or youth. The age of new communicants in the Congregational Church during this period, for example, tended to fall.[34] Adolescents more than any other group are likely to have psychological disorders when suffering from ergotism.

The wide dimensions of the epidemic may have been related to the age structure of the population. In 1741 the proportion of persons under the age of twenty-one may have been high. According to my calculations, based on tax lists and genealogies, in Groton, Connecticut, the average number of children per household having children in a given year was:

1667	3.73
1705	3.62
1717	4.00
1730	3.63
1741	4.23

Given the high susceptibility of the young to ergot poisoning, the at-risk population was unusually large in 1741 in Groton, a major center of awakening. (Fertility was unusually high in colonial New England, which may lead one to ask why, if ergot poisoning was endemic, fertility was not instead depressed. The answer is that the Claviceps strains in New England, like those in Russia, probably did not produce fertility-suppressing alkaloids.)

If ergotism was prevalent in 1741, why wasn't it diagnosed? One reason may be that, although the disease was recognized in Germany at this time, it was not described in English-language publications

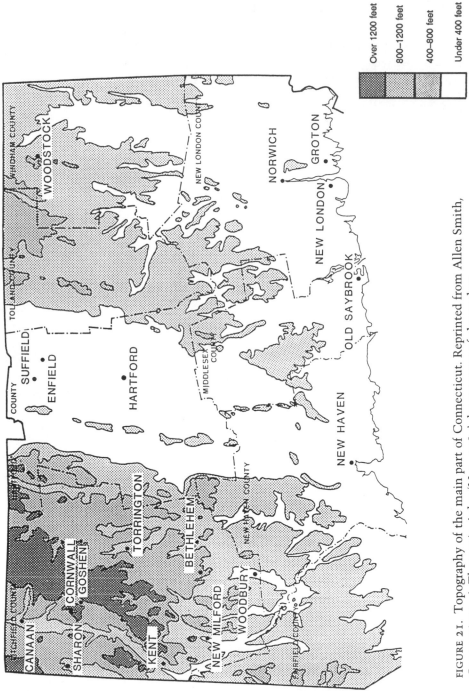

FIGURE 21. Topography of the main part of Connecticut. Reprinted from Allen Smith, *Connecticut: A Thematic Atlas* (N.p., 1984), by courtesy of the author.

Over 1200 feet

800–1200 feet

400–800 feet

Under 400 feet

until 1765. Nevertheless, during the epidemic a few individuals perceived the symptoms as those of a "nervous disorder" or other

TABLE 11. Admissions to the Congregational churches of Connecticut, 1735–1743

Church	1735	1736	1737	1738	1739	1740	1741	1742	1743
Ashford	13	3	4	4	12	13	22	38	0
Bethlehem						4	8	3	7
Bloomfield					9	2	13	10	5
Bozrah					12	4	23	13	2
Branford	5	5	3	11	6	4	10	3	5
Canterbury	5	10	14	9	8	9	3	7	2
Chatham	12	9	6	2	3	3	13	18	3
Cheshire	23	11	12	7	7	15	50	24	1
Colchester	19	33	15	11	0	3	28	37	5
Cromwell	5	0	4	5	15	14	9	16	5
E. Haddam	5	4	4	8	3	12	3	13	0
Fairfield	6	2	9	3	4	7	13	15	2
Franklin	14	21	8	2	16	10	68	33	5
Goshen (Lebanon)	13	21	11	19	7	3	25	41	3
Greenfd Hill (Fairfield)	8	7	5	9	5	10	5	16	4
Griswold	13	15	11	7	10	1	9	46	9
Groton	10	6	2	0	2	1	80	33	6
Hampton	13	12	6	7	8	8	53	66	5
Hartford	4	5	5	16	3	3	27	5	4
Kent							40	7	9
Killingly							60		
Killingsworth				12	17	3	12	10	4
Lebanon	28	20	28	13	6	10	95	52	9
Lyme							113		
Mansfield	85	19	8	8	7	9	31	20	0
Milford	28	23	4	49	8	2	71	8	1
Lisbon							30	5	
New Canaan	11	10	9	1	4	4	4	12	
New Hartford							3	4	5	1
New Haven	45	56	36	20	13	10	28	8	7
New London	19	10	5	16	2	4	84	12	4
New Milford	11	8	7	8	16	7	22	7	3
N. Stonington	10	4	2	13	10	9	15	99	2
Norwich	4	9	6	4	1	0	32	44	15
Old Lyme	13	6	3	8	7	8	150	11	11
Old Saybrook		6	14	11	14	15	5	2	4
Pomfret							106		
Preston	6	8	4	6	9	8	14	2	0

Church	1735	1736	1737	1738	1739	1740	1741	1742	1743
Scotland							49		
Somers	27	4	6	2	1	8	29	38	2
Stonington	13	10	4	5	2	2	16	19	2
Southington	5	18	9	0	3	3	28	11	4
Stratford	31	15	2	1	12	16	7	24	31
Suffield	120	13	4	6	15	5	176	42	15
Trumbull	6	13	2	3	1	3	15	2	3
Voluntown*	9	10	9	5	0	4		1	
W. Hartford	30	4	5	5	7	8	46	7	0
Wethersfield	0	0	0	0	8	5	43	24	4
Windham	21	25	20	17	13		69	31	14
Woodbury	13	10	13	14	33	19	45	40	8
Woodstock	3	0	0	26	13	25	16	11	15

Sources: Original church records deposited in the Connecticut State Library, Hartford; Ellen D. Larned, History of Windham County (Worcester, Mass., 1874–1880), vol. 1, pp. 444 and 450; Thomas Prince, ed., The Christian history (Boston, 1744–45), vol. 2, p. 109; D. H. Hurd, History of New London County (Philadelphia, 1882), p. 732.

*Voluntown was affected by the Great Awakening, but since a separatist movement developed early there, admissions to the established church did not increase. See Larned, History of Windham County, vol. 1, p. 457.

sickness. The clergy were at first divided among themselves as to whether they were confronted with true conversions or illness. Only after the "converts" began to show lack of respect for the authority of the ordained clergy, and to form separate churches, did the majority of clerics come down against revivalism, arguing that the "conversions" were not inspired by God.[35]

BEFORE AND AFTER THE GREAT AWAKENING

The abnormal symptoms that appeared in the New World in 1741 were not truly novel. In colonial annals one often finds references to epidemic symptoms of the central nervous system.

John Winthrop, Sr., said that in 1636 some members of the congregation in Dorchester, Massachusetts, had "dreams and ravishes of spirit by fits." It was a time of food scarcity, following a cold winter. Like the English sectarians, "they expected to believe by some power of their own, and not only and wholly from Christ," commented Winthrop with disapproval.[36]

Robert Calef reported that in the winter of 1692–1693, following the witch trials, a religious revival in Salem was marked by the same central nervous system symptoms. On September 20, 1693, Margaret

Rule, a young woman, suffered for six weeks with fits, hallucinations, blisters on her skin, difficulty swallowing, and a ravenous appetite.[37]

An English physician in Boston wrote in his commonplace book in 1711 about "hysterick fits" he had observed.[38] The symptoms included spasms, violent pain, vomiting, diarrhea, watery swellings on the limbs, coldness of the extremities, depression, laughing and crying without cause. "And I, T. Robie," he wrote, "once heard an hysterick woman pray very zealously, & break off very abruptly and then go to singing; by & by to dancing, etc." Patients of both sexes suffered from spasms accompanied by palpitations of the heart and insomnia: they were "afraid to sleep, as soon as ever they take a Nap, they presently start at 1000 phantasms, & are forced to keep waking."

Epidemics of fits were common in the Connecticut River Valley, which was a frontier region in the first half of the eighteenth century, and a lowland area as well. On February 22, 1716, Stephen Williams reported that in East Windsor there was an extraordinary stir, with people crying out, "What shall we do to be saved?" On July 11, 1716, Williams said that he had "ague" and had been given to making "fearful orrations" and prophesying. On July 18 he wrote, "We hear great talk about witchcraft." On the twenty-eighth he had his "worst fit ever." On August 1, "feeling much amiss," he saw, as did others, a terrifying "flame of fire" just after sunset in the southeast. His later notations included:

> JUNE 14 [1735]. I am fearful of ye palsy. I have a tremour sometimes in my limbs.
> JULY 8. Poor N. Burt 2nd in considerable disorder.
> JULY 11. I find myself considerably discomposed and disordered—full of notions.
> JULY 13. N. Burt 2nd cut his own throat.
> OCT. 21. Saml Chandler child taken with C-fitts.
> NOV. 16. My wife seems considerably disordered. "vapoury disorders"?
> JUNE 17 [1738]. My son John but poorly—has had a fitt something like ye Epilepsie.
> JUNE 26. Family, esp. wife, doing poorly.
> JUNE 30. I—headache, dizziness
> JULY 8. Last night my neighbors reaped my field of rye.

In 1736, the famous preacher Reverend Jonathan Edwards, a neighbor of Williams in the valley, wrote,

There have been many instances in this and some neighbor towns, before now, of persons fainting with joyful discoveries made to their souls, once several together in this town. And there also formerly have been several instances here of persons' flesh waxing cold and benumbed, and their hands clenched, yea, their bodies being set into convulsions with a strong sense of the astonishing great and excellent things of God and the eternal world.[39]

William Douglass, a physician in the Boston area, did not give such symptoms a religious interpretation. Remarking that they were the same as those that had appeared in Salem in 1692, he added that they were endemic in Salem and Ipswich "to this day" as "Hypochondrick, Hysterick, and other Maniack Disorders."[40] The opinion of Douglass is of particular value, as he was the only doctor in New England who was a graduate of a medical school. (Other physicians were trained by apprenticeship.)

In the fall of 1740, the year prior to the Great Awakening, symptoms of dystonic ergotism were present, but to a lesser degree. In the fall, during the tour of Reverend George Whitefield through New England, some cases of conversion occurred. Whitefield was sick during this tour: he felt weak in Boston on September 22, had a vomiting fit on September 28, and again felt weak October 22.[41] The wife of Reverend Ebenezer Parkman in Westboro had a terrible fit on October 10.[42] A pamphlet by Benjamin Lord, written in 1744, described a case in detail: that of Mercy Wheeler of Plainfield, Connecticut, a victim of nervous fever at the age of twenty (in the summer of 1726) who had recovered partially, after several relapses, by 1743. Between attacks she was ravenously hungry.[43]

In the fall of 1745 "nervous fever" was epidemic in Salem, and Dr. Edward Holyoke made note that Mr. Appleton was ill with "St. Anthony's Fire" (a common English term for gangrenous ergotism).[44]

In Waterbury, Connecticut, at harvest time in 1749, a lethal epidemic of "nervous fever" struck the community.[45] A similar but less severe attack was noted in the same year in New London.[46] Stephen Williams reported it in Longmeadow in August 1763 and August 1773. But in the latter part of the eighteenth century the disease grew less common in New England.[47] It lasted longer in the frontier regions of Virginia and Kentucky, but here too the symptoms became rare in the 1790s.[48]

One reason for the declining incidence of "nervous fever" may have been *declining consumption of ergotized rye.* More wheat was marketed. The longer land was under cultivation, the less likely the rye grown on it would get infected.

Later, though, a shift occurred from the cultivation of wheat back to rye. Beginning in 1785 a new pest, the Hessian fly, had begun attacking wheat crops in New Jersey and the Connecticut River Valley. By 1788 this pest reached eastern Pennsylvania; in 1797 it was west of the Allegheny Mountains and by 1800 it had spread throughout Connecticut and entered backcountry New England and New York.[49] Soon the symptoms of ergotism *reappeared* in New England. In Salem, Massachusetts, thirty-three people died of "nervous fever" and of "convulsions" during 1795.[50] Doctors described similar epidemics in Suffield (1811), Lyme (1812–1813) and Mansfield (1813).[51]

After the middle of the eighteenth century the symptoms described here were interpreted naturalistically, as a disease, in New England. On the Kentucky frontier, however, it was otherwise. In the summer of 1800 many Kentucky pioneers who attended camp meetings had fits and hallucinations. They also trembled, jerked, barked, danced compulsively, and fell into trances. In the summer of 1801 this happened again; lesser episodes occurred over the next two years. According to Peter Cartwright, a Methodist preacher:

> From these wild exercises another great evil arose from the heated and wild imagination of some. They professed to fall into trances and see visions, and lay apparently powerless and motionless for days, sometimes a week at a time, without food or drink; and when they came to they professed to have seen heaven and hell, to have seen God, angels, the devil, and the damned; they would prophesy, and, under the pretence of Divine inspiration, predict the time of the end of the world, and the ushering in of the great millennium.[52]

As a premonition, victims experienced a prickly sensation, as if the hands and feet were asleep (this symptom is called formication) and felt a weight on the chest or about the heart. In nearly all cases the pulse was weak and low and the bodily extremities cold (a result of vasoconstriction). The muscles of some were "corded as in nervous complaints" and others had difficulty breathing. Some people experienced ecstasy, while others thought they deserved Hell. Some felt the compulsion to laugh and speak incessantly, while others suffered loss of speech. Some wept or shrieked before they fell unconscious.

The fits might last half an hour to an hour. They might recur, with remission, over many weeks and months and even years.

These episodes surprised many observers. The frontier people at this time were generally indifferent to religion, and they were not given to emotionalism of any kind. In some instances a preacher could stop the spread of the symptoms in a camp meeting, and so it had been theorized that the symptoms were psychological in origin. But it may just as well have been the case that the audience was highly suggestible. Moreover, people also had fits outside any religious context: in a meeting house, cabin, field, school, or at the side of a road. They might develop the symptoms both alone and in company. Those affected were between eighteen months and seventy years of age, but children and young people were more likely to be affected than full-grown adults.[53]

The earliest and most severe symptoms were found among the Presbyterians, a group consisting mainly of Scots and Scotch-Irish (from Ulster). In the eighteenth and early nineteenth centuries these people were still accustomed to eating rye bread, and to this end they grew small quantities of rye for domestic use.[54] As a result of population pressure at home, the pace of emigration of the Scots to America increased after 1763; around 1800 they constituted about one-third of the population of the frontier settlements of Kentucky and Tennessee.[55] Not only for the Scots, but for the frontier settlers in general, a dietary staple around 1800 was mush made of cornmeal or rye flour and consumed for supper by both adults and children.[56]

Of all the localities in which the symptoms appeared, not one was at a higher elevation than 850 feet above sea level.[57] The timing was also appropriate: the symptoms appeared with sudden severity in the last week of July or early August.[58] With the onset of winter, they tended to dwindle. Lesser outbreaks occurred during 1801–1803.

It is somewhat puzzling that the German population in the middle colonies, among whom rye bread was popular, did not experience an epidemic of ergotism symptoms in these years.[59] The answer may lie in the fact that the Germans knew that ergotized rye was dangerous, especially after the epidemic of 1770–1771 in Germany. Most likely German midwives knew many of the properties of ergot long before; one midwife in upstate New York in 1807 informed Dr. John Stearns that ergot could speed labor and prevent bleeding in childbirth.[60] In addition to the Germans, there were other farmers in New York who grew rye, but instead of consuming it they sold it to whiskey distillers.[61] Ergot alkaloids do not persist in distilled alcoholic drinks. So while there were religious revivals in upstate New York in the early nineteenth century, they were not characterized by the violent symptoms that appeared on the Kentucky frontier.[62]

For the historian, an understanding of epidemic central nervous system symptoms may illuminate the conflict between ministers and physicians for leadership in their communities. Not all ministers saw such symptoms as evidence of divine intervention and conversion, but many were carried away with enthusiasm for the harvest of souls that sometimes resulted. This was more likely to occur where churches were not yet well organized and financed; such communities might be induced to offer a minister a secure position. When ministers realized that "excitement" was destabilizing for the community, that it just as easily served the purposes of rebels against established leadership, they joined with physicians in interpreting the symptom naturalistically, as those of disease. In this way they delegitimized the "converted." Unfortunately, frontier physicians apparently knew neither the cause nor the cure of the disease.

In the next chapter I will move from examination of bizarre behavior to social responses to it. As I noted at the beginning of this chapter, there are *two* questions in need of attention in regard to events such as the Great Fear, witch persecution, and the Great Awakening. All along I have been addressing one question: what caused the type of behavior we now label "bizarre"? Just as interesting is the second question: what caused the response to such behavior? And how did this response change over time—and why?

PART IV

Reflections

CHAPTER ELEVEN

Social Control of Mass Psychosis

ILLNESS upsets society in many ways. An epidemic of central nervous system symptoms may disturb existing political arrangements. It may disrupt the ways in which a community controls the behavior of individual members. It may enable some to settle old scores by finding scapegoats, and others to rise in the pecking order by claiming they have divine inspiration.

What happens depends on the screen of interpretation that people adopt. Sometimes an individual chooses a screen for self-serving "reasons." But more often people merely accept the most prevalent screen, the one that is culturally mandated. This cultural mandate changes through time. Here it seems fitting to ask, why is one screen more prevalent than another at a given time? In the past the principal screens of interpretation were:

1. *The negative supernatural interpretation.* The person experiencing nervous disorders was described as being possessed or seized by an evil or mischievous supernatural being—demon or god—or afflicted by the magic of a witch, sorcerer, or the like. The expression, *convulsive seizure,* literally "seizure" by an evil spirit, is a fossil of this way of thinking. If the supposed supernatural being was thought to act alone, the symptoms were treated with some Christian countermagic such as exorcism (see Figure 22). If a witch was believed to be involved she, as the community's scapegoat, might be tortured or killed.

2. *Positive supernatural interpretation.* People who experienced pleasant hallucinations or euphoria, at least part of the time, might claim to be favored by God. On the basis of this claim they might seek to obtain greater influence within the community. If they gained

FIGURE 22. The negative supernatural interpretation: victims of nervous disorders are seen as being under the influence of evil beings. A common response was to drive away the spirit, as is shown in this scene of Saint Ignatius of Loyola exorcising the possessed. (After a painting by Peter Paul Rubens, 1577–1640; photo from the National Library of Medicine.)

wide enough credibility—as seers, prophets, the elect—they might challenge the established power structure (see Figure 23). This was what happened in Westmoreland and Cumberland in the 1650s, when the Quakers left the Anglican Church. The same thing occurred after the Great Awakening of 1741 in New England, when the Separatists split from the Congregational Church. Some such mechanism may have been involved in the history of heretical sects in the medieval period and of some radical Protestant denominations in the Early Modern Era.

3. *The naturalistic interpretation.* The ruling elite might see behavioral disturbances as independent of any supernatural intervention or of human action, magical or otherwise. They might define those who experienced them as sick; the pattern of symptoms, as a disease. They might call the disease epilepsy, hysteria, convulsions, the mother, palsy, vapors, nervous fever, and so on. Then they could treat the person having the symptoms as a victim of a misfortune. At the same time, the victim would not be a hero or martyr and would receive less, not more, respect in the community.

4. *The pragmatic approach.* Those who became the leaders of sects permitted individuals, when the Spirit came upon them, to behave in deviant ways, but not to the point of disrupting the functioning of the group. When such behavior became disruptive they treated the deviants as sick and in some way excluded them from group activities.

Why did one interpretation prevail instead of another? Before attempting an answer, we may find it helpful to imagine the probable consequences of adopting each of the four approaches. If people blamed the symptoms of nervous disorders on demonic possession and the clergy failed to remove them by exorcism, the clergy lost prestige. Singling out a witch or sorcerer for punishment might stabilize the community—but only if the scapegoat had no powerful friends or relatives. Moreover, a witch-hunt might get out of hand and become a general settling of scores between factions in the community. This would cause more trauma.

If those *experiencing* the symptoms gave them a positive supernatural interpretation, they might increase their own authority to the point of challenging the legitimacy of the established authorities. Although religion and politics are not de facto separate today, in the past they were enmeshed to a far greater degree, and anyone who challenged religious authority became directly or indirectly a challenger of secular authority. Before the nineteenth century, there were

FIGURE 23. The positive supernatural interpretation: victims of nervous disorders are seen as being under the influence of divine beings. As a result, religious enthusiasm increased. This engraving, "Agitations des convulsionaires," shows an outbreak of the type common in the seventeenth and eighteenth centuries. Photo by permission of the Wellcome Institute Library, London.

no standing police forces; before the eighteenth century, only a few states had standing armies, and they were small. Most rulers could not afford to maintain large coercive forces, so they defended their position not so much by systematic coercion as by the more subtle influence of the established church.

To set up a sect in opposition to the established church was, therefore, a politically subversive act, whether it was intended to be or not. The political establishment could, and often did, respond by saying that the new sect was heretical and that its self-appointed prophets were minions of the Devil. The established church might use its courts, and the secular authorities their armed forces and courts, to suppress the sect. Or they might construct a new screen of interpretation.

CHRONOLOGY OF DOMINANT INTERPRETATIONS

Contrasting interpretations, often attributable to differences in social and educational status, existed in all periods, but it is possible to identify a dominant influence in certain periods.

1430–1491. During the demographic depression in northern Europe the negative supernatural interpretation prevailed: there was mass witch persecution. Perhaps anxiety was high because the symptoms were very common at this time, and people felt a strong need for a scapegoat. In rural areas the prevailing paganism mandated that folk healers should be the scapegoats.[1]

1492–1560. The climate was warm in most parts of Europe and epidemics of nervous disorders were rare and localized. Cases of alleged demonic possession and bewitchment were also rare. At this time Jean Riolan (1538–1606) and Johann Weyer (1515–1588) articulated the naturalistic interpretation, but they made no clear distinctions between "convulsions," "epilepsy," and "hysteria" and spoke of noxious vapors as a cause of the symptoms.[2]

The positive supernatural interpretation appeared in Germany at this time. During the 1520s, when the weather in Germany was wetter than that in the rest of Europe, the symptoms were probably more prevalent than elsewhere. The winter of 1517 in western Europe was severe, and it was followed by an outbreak of gangrenous ergotism in Strasbourg (upper Rhine Valley). Following the severe winter of 1534 epidemic hallucinations and convulsions appeared in the Anabaptist (radical Protestant) city of Münster. Central nervous system symptoms were most prevalent among Anabaptists in regions that were normally cold and wet and given to rye cultivation: the valleys of the Austrian Tirol, northeastern subalpine Switzerland, the upper Rhine Valley, Swabia, Hesse, West Thuringia, and Moravia. Anabaptism did not thrive in drier East Thuringia and the Elbe River Valley.[3]

Symptoms were not reported in large numbers among mainstream Protestants. The Reformation began at a time when such symptoms were relatively uncommon; indeed, the mainstream Reformation may have become viable in part because it was not excessively "enthusiastic." It is not clear, however, why John Calvin and Ulrich Zwingli, two of the three most important early Protestant leaders, were most successful in Switzerland. A clue may be that this country was dependent on rye crops that were highly vulnerable to ergot infection on account of alpine climatic conditions. Nor has it ever been explained why the Rhine Valley, from Basel and Strasbourg to the Netherlands, was an early focus of radical Protestant activity. Here, too, rye crops were especially prone to ergot infection because the elevation was low, winters were colder and summers wetter than in neighboring areas. Epidemics of "bewitchment" also occurred in the same general areas: the alpine districts and Rhine Valley.

1560–1661. During this period the incidence of epidemic hysterias and plague rose again. Negative supernatural interpretations, especially in the form of witchcraft accusations, were dominant. Positive supernatural and naturalistic interpretations were also present.

Religious enthusiasm in England reached a peak in the middle of the seventeenth century, when many individuals had visions, tremors, and fits and went about foretelling the coming of a new day. Familists, Seekers, Ranters, Muggletonians, and Fifth Monarchy Men sprang up here and there like mushrooms. Even people in orthodox religious groups had millennial expectations by 1649, and 1649–1660 every kind of prophecy was aired, even by young children. "On a half a dozen occasions between 1647 and 1654," reported Keith Thomas, "the deliberations of Oliver Cromwell and his colleagues were interrupted, so that some obscure prophet, often a woman, could be admitted to deliver her message."[4]

The anti-Quaker writer Francis Higginson reported that during Quaker meetings men, but more often women and children, fell into quaking fits. They would faint as though struck with epilepsy or apoplexy and lie, either struggling or quiet, their lips quivering, flesh and joints trembling, belly swollen, foaming at the mouth and "sometimes purge as if they had taken Physick." Such fits lasted one to two hours. Sometimes the Quakers let such afflicted individuals lie; other times they took them home and put them to bed. Higginson claimed that Quakers became "strangely distracted" for prolonged periods in that state.[5] John Gilpin of Kendall, a former Quaker, wrote that he was tempted to cut his own throat, and William Pool of Worcester went insane and drowned himself.[6]

Quaker accounts of their own meetings omit disagreeable details, yet they are revealing nevertheless. For example, at a mass meeting in Bristol on September 10, 1654, Charles Marshall, aged seventeen, reported that "All my limbs smote together and I was like a drunken man because of the Lord and because of the Word of His holiness, and I was made to cry like a woman in travail." Relating an experience at midnight in prison in the spring of 1654, Thomas Holme wrote, "I scarcely knew whether I was in the body, yeah or no, and there appeared light in the prison and astonished me, and I was afraid, and trembled at the appearance of the light, my legs shook under me and my fellow prisoners beheld the light and wondered, and the light was so glorious it dazzled my eyes."[7]

No matter how the symptoms were interpreted, their geographic distribution was within the area that included Scotland, England, France, the Low Countries, Germany, northern Italy, and northern Spain. In Ireland and Wales, where dairy products were very important in the diet and hardly any rye was grown, the symptoms were absent. The symptoms were rare in Scandinavia, where the summers were too cold to favor alkaloid formation. They were uncommon in central and southern Spain and Italy, where wheat and barley were the dominant crops.

Even though tribunals often refused to heed their testimony in courts, physicians continued to interpret central nervous system symptoms naturalistically, attributing them to bodily humors. Among these physicians were Edward Jorden, Jean Taxil, and Charles Le Pois. The Marburg medical faculty accurately described epidemic nervous disorders of 1596 in Hesse and Westphalia.[8]

Symptoms of the nervous system interpreted as those of demonic possession were relatively common from 1560 to 1566. They occurred mainly in the same geographic area as other epidemics of nervous disease. Almost all occurred in convents or orphanages of the Catholic Church. Within the precincts of the church, the clergy treated people suffering with such symptoms by exorcism, but in the epidemic at Loudun a priest became the scapegoat and was executed.[9]

1662–1750. There were spasms of epidemics of hysterias after

1661, but the overall incidence was lower than in the preceding period. After the excited "Lamb's War" of the 1650s the Quakers cooled off. They prophesized more rarely; by 1700 influential English Friends regarded prophesying as odd.[10] In 1664 Richard Baxter wrote of the Quakers:

> At first they did use to fall into violent Tremblings and sometimes Vomitings in their meetings, and pretended to be violently acted upon by the Spirit; but now that is ceased, they only meet, and he that pretendeth to be moved by the Spirit speaketh; and sometimes they say nothing but sit an hour or more in silence, and then depart.[11]

Quaker meetings devised institutional controls for disturbing behavior of members. Indeed, the wild prophesying of the early 1650s served as an incentive to develop such controls, lest the group disintegrate. One was "eldering," the practice of gently escorting a raving individual away from the meeting. Also, according to Quaker teaching, any supposedly divine direction that came to the mind of a Quaker had to withstand three tests before it was followed: Was it moral? Did it stand the test of time? Was it consistent with the Bible and other revelations of the Holy Spirit? Individual Quakers made a practice of submitting their writings to the group for approval before publication.[12]

Meanwhile, the attitude of the British establishment was changing. Because physical persecution of religious deviants tended to produce martyrs, it was counterproductive as a social control device. Establishment writers, drawing upon ancient medical sources such as Hippocrates and Celsus, labeled disturbances of the central nervous system as hysteria, epilepsy, cholera morbus, frenzy, and nervous fever. This approach did not physically traumatize anyone or make martyrs; instead, it served to discredit and delegitimize religious rebels against authority, who were depicted as sick, incompetent, pitiable.[13]

Sixteen eighty-eight, when the climate was favorable to ergot alkaloid formation, was a year of transition in attitudes toward central nervous system symptoms among the elite of western Europe. In the Dauphiné (France) several hundred children showed signs of nervous disorder, and their neighbors gave them positive supernatural significance. This was the beginning of the Camisard movement, which later became politically subversive. The French government was skeptical.[14] In England, a preacher reported that at a Quaker meeting in 1688 there was an "outbreak of the Lord's Prayer" that attracted the attention of non-Quakers for fourteen miles around. Also, a panic occurred in Cambridge, Nottingham, the Midlands, and farther north, beginning September 30. People thought the disbanded Irish troops of James II were coming to burn and loot their towns. In all cases the elite were skeptical.[15]

By the early eighteenth century western elites doubted any supernatural interpretation of behavioral disturbances. In London they derided and pilloried the Camisard émigrés; in Paris they challenged the convulsionaries of St. Medard.[16]

1750–1850. Winters were colder in this period, but in some areas rye consumption declined. This was a time of increased reporting of epidemics of nervous disease. The negative supernatural interpretation disappeared altogether, as did witch persecution. The positive supernatural interpretation went out of style. To be sure, several hundred people had visions in Bristol on July 13, 1741, and also in Cambuslang parish, Lanarkshire, on February 18, 1742. Similar outbreaks occurred in Everton, West Riding, and Yorkshire in the summers of 1758 and 1759, in the Shetland Islands in the 1760s, in Redruth, Cornwall, in 1841, and in Småland, southern Sweden, in 1841, but they were described in the press as inexplicable but natural events.[17]

Meanwhile, in Germany the great ergotism epidemic of 1770–1771 prompted so much investigation that the naturalistic interpretation was put on a sound scientific basis. Gradually this scientific information spread among European physicians.

During the nineteenth century, as rye consumption declined, epidemics of spasms, tremors, chorea, panic, and hallucinations also became rare. Medical textbook writers gave ergotism little attention. So obscure did this disease become that in 1976, when Linnda Caporael published an article linking the disease with the Salem witchcraft affair, most educated people had never heard of it.[18]

Today our food supply seems safe enough that we rarely consider food poisoning as a possible cause of bizarre behavior. We no longer need food poisoning as an explanation, but such an explanation may be highly appropriate for past bizarre behavior. This is why I have called attention to forgotten and overlooked forms of food poisoning.

Such historical research seems justified in light of current brain research. The biochemistry of the brain is no longer a complete mystery. Some psychiatrists go so far as to say that "for every twisted thought there is a twisted molecule."[19] This may not turn out to be true, but it seems reasonable that what poisons the brain, poisons the mind.

This being so, a new door is opening to the understanding of past mentalities. For example, some day it may become possible to comprehend the reasons for genocide, war, and other forms of violence. Theories that originate in the laboratory may then be tested against patterns of past events.

CHAPTER TWELVE

Plant Health and Human Health

IT may seem irregular to use plant health, and climatic indicators of plant health, as an index of human health, but the two are intimately related. If the plants that people eat are toxic or diseased, people will be either diseased or simply not as healthy, as robust and fertile, as they would be if they were properly nourished. Just as an individual's vigor will diminish if poisoned, so will a population lose its ability to thrive if its food supply is tainted. The population in the area identified more or less with the heart of Western civilization did not begin to thrive until the sixteenth century (and then only slowly until about 1750), and they grew at different rates in different regions. These regional differences, in fact, highlight the effects of plant health on human health because of the variable conditions for plant growth in different environments. Consider the following overview of western Europe at the beginning of its period of population increase.

For two hundred years after the Black Death Europeans north of the Alps and Pyrenees struggled to avert extinction. By the 1490s they had recovered from their losses, but still population levels made only small gains.

A fundamental fact of life at this time was the dependence of most of the population on rye bread. Rye could be grown in a cold climate and in sandy and sour soils, but as human food it had two severe drawbacks: it was unique, among common cereal grains, in being vulnerable to infection by the fungus *Claviceps purpurea*; and it was the best substrate for the development of *Fusarium sporotrichiella* strains capable of producing lethal trichothecene toxins. Europeans were not oblivious to the dangers of eating rye. Even in Roman Gaul, the ruling class preferred wheat products to those made of rye. Whenever wheat products were available, those who could afford them bought them, keeping the price high. This implied that people were to some degree conscious that eating rye products was undesirable, even though they did not know precisely why. Those who could eat wheat, did; and in time of dearth they were fewer in number than in time of plenty. When New World crops entered Europe the supply of safe staple starches expanded. I believe that this was the most important single cause for health improvement prior to the twentieth century.

As noted before, different strains of *Claviceps* produce different amounts of toxic substances, and this is a matter of genetic variation. In general, lethal mycotoxins tend to form on cereals grown in low, moist places. In addition, ergot is most abundant when May and July temperatures are in the 17.4°–18.9° C range. The areas of Europe that fit this climatic pattern most closely were central and eastern Russia and the high valleys of the Alpine areas.

Russia was the most ill-favored country of Europe from the point of view of mold poisoning. Especially ill-favored was Muscovite Russia, where the *Claviceps* strains produced ergots with a high alkaloid content. The cold climate forced dependence on rye as a staple; the cold winters traumatized the fall-sown rye seedlings, making them susceptible to *Claviceps* infection; and the summer temperatures maximized alkaloid production. There was only one natural advantage for human beings living in Muscovite Russia: the content of fertility-suppressing alkaloids was apparently low. The Russians benefited from this feature of local *Claviceps* strain: their high infant mortality rate was offset by early marriage and rapid-fire conceptions.

In general, the further west one moved from Russia across the European continent, the less danger one was in of dying from ergot poisoning. But ergot alkaloids probably had a strong influence on fertility even in the far western areas. Wheat could be grown in loamy soils deposited by the great rivers flowing from the mountains to the sea. These same waterways facilitated commerce in grain—and spread the effects of food poisoning when crops were infected. In continental western Europe white wheat bread was the staple food of the ruling class.

The coastal areas and islands of northwestern Europe had a maritime climate. Consequently they had neither severe winters nor summer temperatures warm enough for maximal *Claviceps* infection of rye. It is likely, however, that the prevalent *Claviceps* strains had a high content of fertility-suppressing alkaloids and a low level of the most lethal alkaloids.

The moistness of the climate in much of northwestern Europe favored the formation of *Fusarium* toxins in carelessly stored cereals. In countries like England, where mortality usually peaked in the spring, *Fusarium* toxins were probably an important underlying cause of death.

Although in colonial New England the British settlers immediately adopted a new crop, maize, which from the beginning was the most important starch (about three-fifths of the total diet), they did not escape food poisoning. They brought with them wheat, rye, oat, and barley seeds to try out in the new environment. Wheat was subject to "rust" (infection by *Puccinia graminis*) and did not thrive in the sandy soils and cold climate of northern New England, so during the eighteenth century the settlers grew rye—about two-fifths of the total starch in the diet. Rye consumption did not constrain fertility in New England, but it appears that ergotism was endemic there and was known as "nervous fever" or "fits."

In the sixteenth century new crops (maize and potatoes) from the New World began to enter Europe, which apparently improved health. In addition, spells of warm weather stimulated population growth. One such spell occurred during 1500–1560. As the growing population demanded more food, farmers used more labor-intensive agricultural techniques. They drained swamps, sweetened sour soils with lime, and planted maize, potatoes, and other New World crops.

In the period 1560–1661 colder weather and the resurgence of plague and ergotism caused setbacks in population growth. Witch persecution flared up, disturbing rural communities, and religious wars disrupted the grain trade.

After 1661 the weather improved, governments were better prepared for famines, and population stabilized. In England improved techniques increased wheat production, and in Ireland the population thrived on a diet of potatoes and milk. In the British Isles, a population surplus, so important for the Industrial Revolution, began to accumulate by the middle of the eighteenth century.

In Russia population increased not by increasing productivity, but by conquest and colonization. In the seventeenth century the population of Muscovite Russia moved southward into the Ukraine, opening up the black soil belt to cultivation. Here the *Claviceps* strains were relatively harmless. Improvement was most dramatic and rapid in the eighteenth century.

During the nineteenth century the potato became a staple food of the masses in most parts of western, central, and northern Europe. People also learned about the dangers of ergot. In the second half of the century western and central Europeans began to import cheap wheat from the New World, to clean up the water supply, to prosecute food adulterers, and to sanitize hospitals. The Russians built railroads, facilitating famine relief in remote areas. After 1850 they, too, began to consume substantial quantities of potatoes. Diversification of crops increased. Availability of a variety of staple starches not only was a hedge against famine, but reduced the probability of severe infection of any one food plant.

Trends in plant health have apparently influenced trends in human health in Europe and America. As a consequence, one can learn much about demographic history by studying agricultural history. Man does not live by bread alone—but he must have bread. And he must have bread that is truly a staff of life, not a scepter of death.

Notes

1. FOOD POISONING AND HISTORY

1. Nancy Birdsall, "Fertility and economic change in eighteenth and nineteenth century Europe: A comment," *Population and Development Review* 9 (1983), pp. 111–135; Stephen J. Kunitz, "Speculation on the European mortality decline," *Economic History Review*, 2d ser., 36 (1983), pp. 349–364; Peter H. Lindert, "English living standards, population growth, and Wrigley-Schofield," *Exploration in Economic History* 20 (1983), pp. 131–155; Roger Schofield, "The impact of scarcity and plenty on population change in England, 1541–1871," *Journal of Interdisciplinary History* 14 (1983), pp. 265–291; E. A. Wrigley, "The growth of population in eighteenth century England: A conundrum resolved," *Past and Present* 98 (1983), pp. 121–150; David R. Weir, "Life under pressure: France and England, 1670–1870," *Journal of Economic History* 44 (1984), pp. 27–47; Myron F. Gutmann and Rene Leboutte, "Rethinking protoindustrialization and the family," *Journal of Interdisciplinary History* 14 (1984), pp. 587–609; Mary K. Matossian, "Mold poisoning and population growth in England and France, 1750–1850," *Journal of Economic History* 44 (1984), pp. 669–686.

2. Robert D. Lee, ed., *European demography and economic growth* (New York, 1979).

3. R. D. Lee, "Population in pre-industrial England: An econometric analysis," *Quarterly Journal of Economics* 87 (1973), pp. 581–607.

4. Ronald Lee, "Population homeostasis and English demographic history," *Journal of Interdisciplinary History* 15 (1985), pp. 635–660.

5. Fernand Braudel, *The Mediterranean and the Mediterranean world in the age of Philip II*, vol. 2 (New York, 1973), p. 1241.

6. Susan Cotts Watkins and E. van der Walle, "Nutrition, mortality, and population size: Malthus' court of last resort," *Journal of Interdisciplinary History* 41 (1983), pp. 205–227; John Bongaarts, "Does malnutrition affect fecundity? A summary of the evidence," *Science* 208 (May 9, 1980), pp. 564–569; Jane Menken et al., "The nutrition fertility link: An evaluation of the evidence," *Journal of Interdisciplinary History* 11 (1981), pp. 425–

441; M. J. Murray et al., "Adverse effects of normal nutrients and foods on host resistance to disease," *Adverse effects of foods,* ed. E. F. P. Jeliffe and D. B. Jeliffe (New York, 1982), pp. 313–321.

7. J. B. Peacock and R. J. F. H. Pinsent, "Reported mortality and weather," *Journal of the Royal College of General Practitioners* 25 (1975), p. 250; S. W. Tromp, "Influence of weather and climate on infectious diseases in man," *Progress in Biometeorology* (Amsterdam, 1977), A1, pt. 2, pp. 66–67.

8. Maurice Aymard, "Toward the history of nutrition: Some methodological remarks," in *Food and drink in history,* ed. R. Forster and O. Ranum (Baltimore, 1979), pp. 1–16; T. C. Barker et al., *Our changing fare: Two hundred years of British food habits* (London, 1966); John Burnett, *Plenty and want: A social history of diet in England from 1815 to the present day,* 2d ed. (London, 1979); Hugh Clout, *Agriculture in France on the eve of the railway age* (London, 1980); D. J. Oddy, "Working class diets in late nineteenth century Britain," *'Economic History Review* 23 (1970), pp. 314–323; Oddy, "Food in the nineteenth century: Nutrition in the first urban society," *Proceedings of the Nutrition Society* 29 (1970), pp. 150–157; Oddy, *The making of the modern British diet* (Totowa, N.J., 1976); Jean-Claude Toutain, *La consommation alimentaire en France en 1789 à 1964* (Geneva, 1971); Anthony S. Wohl, *Endangered lives: Public health in Victorian Britain* (Cambridge, Mass., 1983).

9. John Hathcock, ed., *Nutritional toxicology* (New York, 1982); John Lacey, "Understanding the fungal threat to food," *Food and Climate Review* (Boulder, Colo., 1982–1983), pp. 30–41; Claude Moreau and Maurice Moss, *Moulds, toxins, and food* (New York, 1979).

10. C. F. Mayer, "Endemic panmyelotoxicosis in the Russian grain belt," *Military Surgeon* 113 (1953), pp. 173–189 and 295–315; C. C. Hsia et al., "Proliferative and cytotoxic effects of *Fusarium* T-2 toxin on cultured human fetal esophagus," *Carcinogens* 4 (1983), pp. 1101–1107.

11. C. A. Linsell, "The mycotoxins and human health hazards," *Annales de Nutrition et Alimentation* 31 (1977), pp. 996–1004.

12. Pat B. Hamilton, "Fallacies in our understanding of mycotoxins," *Journal of Food Production* 41 (1978), p. 404.

13. B. Berde and H. O. Schild, eds., *Ergot alkaloids and related compounds* (New York, 1978); P. F. Spano and M. Trabucchi, eds., *Ergot alkaloids,* Supplement 1, *Pharmacology* 17 (1978), pp. 1–216; Klaus Lorenz, "Ergot on cereal grains," *CRC critical reviews in food science and nutrition* 11 (1979), pp. 311–354; B. Berde, "Ergot compounds: A synopsis," in *Ergot compounds and brain function,* ed. M. Goldstein et al. (New York, 1980), pp. 3–24.

14. Fritz Siemens, "Psychosen bei Ergotismus," *Archiv für Psychiatrie und Nervenkrankheiten* 11, no. 1–2 (1880), pp. 366–390.

15. Richard E. Schultes and Albert Hofmann, *The botany and chemistry of hallucinogens,* 2d ed. (Springfield, Ill., 1980), p. 37.

16. Ralph Emerson, "Mycological relevance in the nineteen seventies," *Transactions of the British Mycological Society* 60 (1973), pp. 363–387.

17. N. N. Reformatskii, *Dushevnoe rasstroistvo pri otravlenii sporinei (bolezn' zlaia korcha)* (Moscow, 1893), pp. 96, 102, and 365–368.

18. G. A. Kolosov [Kolossow], "Geistesstörungen bei Ergotismus," *Archiv für Psychiatrie und Nervenkrankheiten* 53 (1914), pp. 1118–1129, trans. Bayard Holmes and W. H. Sole in *Chicago Medical Recorder* 38 (1916), pp. 271–279.

19. Abraham Grunfeld, "Kurzer Auszug aus die Mutterkornfrage betreffenden Arbeiten der russischen Literatur," *Historische Studien aus dem pharmakologischen Institut der Kaiserlichen Universität* (Dorpat, 1889).

20. Claude Moreau, "Les mycotoxins neurotropes de l'Aspergillus fumigatus," *Bulletin de la Société Mycologique de France* 98, no. 3 (1982), pp. 261–273.

21. N. Zinin et al., *Izsledovanie o sporyn'e (Secale cornutum) o sposobakh otkrytiia eia v muke* (St. Petersburg, 1864), p. 29; S. F. Popov, "Raspredelenie sporyn'i 1926 g. v Predural'i i sposoby ee opredeleniia," *Gigiena i Epidemiologiia* 8 (1927), pp. 32–40.

22. S. V. Vladimirskii, "Geograficheskoe rasprostranenie i zony vredonosnogo znacheniia sporyn'i na rzhi SSSR," *Sovetskaia Botanika* 5 (1939), p. 82; B. Kosir et al. [Factors affecting the yield and quantity of sclerotia from *Claviceps purpurea*] [in Slovenian, with an English abstract, *Farmatsevtskii Vestnik* 32 (1981), pp. 21–25.

23. L. L. Klaf et al., *Ergotizm* (Kharkov, 1933), p. 10.

24. E. A. Wrigley and R. S. Schofield, *The population history of England, 1541–1871* (Cambridge, Mass., 1981), p. 360.

25. Mayer, "Endemic panmyelotoxicosis"; E. B. Smalley et al., "Mycotoxicoses associated with moldy corn," in *Toxic microorganisms, mycotoxins, botulism*, ed. Mendel Herzberg (Washington, D.C., 1970), pp. 163–173; N. R. Kosuri et al., "Toxicologic studies of Fusarium tricinctum (Corda) Snyder et Hansen from moldy corn," *American Journal of Veterinary Research* 32 (1971), pp. 1843–1850; A. N. Leonov, "Current view of the chemical nature of factors responsible for alimentary toxic aleukiia," in *Mycotoxins in human and animal health*, ed. Joseph V. Rodricks, C. W. Hesseltine, and M. A. Mehlman (Park Forest South, Ill., 1977), pp. 323–328; Moreau and Moss, *Moulds, toxins, and food.*

26. V. V. Rukhliada, "Biosintez T-2 toksina gribom Fusarium Sporotrichiella Bilai na razlichnykh rastitel'nykh substratakh," *Mikologiia i Fitopatologiia* 18, no. 1 (1984), pp. 73–75.

27. Z. I. Malkin, ed., *Septicheskaia Angina i eë Lechenie* (Kazan, 1945), p. 61; B. Chilikin and N. Geminov, "Epidemiologiia alimentarno-toksicheskoi aleikii," *Alimentarno-toksicheskaia Aleikiia* (Kuibyshev, 1945), p. 4.

28. A. I. Nesterov et al., *Alimentarno-toksicheskaia Aleikiia* (Moscow, 1945), p. 8; V. V. Efremov, *Alimentarno-toksicheskaia Aleikiia* (Moscow, 1945).

29. A. I. Stolmakov and A. Y. Nomofilov, "Epidemiologicheskie nabliudeniia nad vspyshkoi alimentarno-toksicheskaia aleikiia v raione B," in *Alimentarno-toksicheskaia Aleikiia*, ed. N. I. Malov et al. (Chkalov [Orenburg], 1947), pp. 8–10.

30. Hsia et al., "Proliferative and cytotoxic effects."

31. Moreau and Moss, *Moulds, toxins, and food*, p. 223; Malkin, *Septicheskaia Angina*, p. 63; A. Z. Joffe, "Toxicity of *Fusarium poae* and *F.*

sporotrichiodes and its relation to alimentary toxic aleikiia," in *Mycotoxins*, ed. I. F. Purchase (Amsterdam, 1974), pp. 229–262.

32. A. Z. Joffe, "Environmental conditions favorable to *Fusarium* toxin formation causing serious outbreaks in animals and man," *Veterinary Research Communications* 7 (1983), pp. 187–193; C. J. Rabie et al., "T-2 toxin production by *Fusarium acuminatum* isolated from oats and barley," *Applied Environmental Microbiology* 52 (1986), pp. 594–596.

33. Y. Rosenstein et al., "Effects of *Fusarium* toxins, T-2 toxin, and diacetoxyscirpenol on murine T-independent immune responses," *Immunology* 44 (1981), pp. 555–560; V. Jagadeesan et al., "Immune studies in T-2 toxin: effect of feeding and withdrawal in monkeys," *Food and Chemical Toxicology* 20 (1982), pp. 83–87; E. Masuda et al., "Immunosuppressive effect of trichothecene toxin, Fusarenon-X, in mice," *Immunology* 45 (1982), pp. 743–749.

34. M. A. Hayes, "Comparative toxicity of dietary T-2 toxin in rats and mice," *Journal of Applied Toxicology* 2 (1982), pp. 207–212.

35. A. C. Pier and M. E. McLoughlin, "Mycotoxic suppression of immunity," in *Trichothecenes and other mycotoxins*, ed. J. Lacey (New York, 1985), pp. 507–519.

36. Regina Schoental, "The role of mycotoxins in certain idiopathic disorders and tumours in animals and man,' in *Nutrition and killer diseases*, ed. John Rose (Park Ridge, N.J., 1982), pp. 139–147.

37. Basil Jarvis, "The occurrence of mycotoxins in UK foods," *Food Technology in Australia* 34 (1982), pp. 508–514.

38. A. Bottalico et al., "Production of zearalenone, trichothecene, and moniliformin by *Fusarium* species from cereal in Italy," in *Toxigenic fungi*, ed. H. Kurata and Y. Ueno (Amsterdam, 1984), pp. 199–208.

39. Gabriella Sándor, "Occurrence of mycotoxins in feeds, animal organs, and secretions," *Acta Veterinaria Hungarica* 32, no. 1–2 (1984), pp. 57–69.

40. B. Gedek and J. Bauer, "Trichothecene problems in the Federal Republic of Germany," *Developments in Food Science* 4 (1983), pp. 301–307.

41. Alan C. Pier, "Mycotoxins and animal health," *Advances in Veterinary Science and Comparative Medicine* 25 (1981), pp. 186–243.

2. A CASE STUDY: RUSSIA AND ITS NEIGHBORS

1. I. V. Stefanovich-Dontsov, *Opisanie o chërnykh rozhkakh, prichiniaiushchikh korchi i pomertevnie chlenov* (St. Petersburg, 1797).

2. Anon., "O khlebnykh rasteniiakh," *Zhurnal Ministerstva Vnutrennykh Del* 41, no. 8 (1841), pp. 235–241.

3. S. K. Bogoiavlenskii, *Drevnee russkoe vrachevanie v XI-XVII vv., istochniki dlia istorii russkoi meditsiny* (Moscow, 1960), pp. 176–177 and 257.

4. *Derevenskoe Zerkalo*, vol. 2 (1798–99), pp. 39–44.

5. Iasiukovich, "Opisanie epidemicheskoe bolezni (Raphania) byvshei v mestechke Vyshkakh v 1824-m gody," *Voenno-Meditsinskii Zhurnal* 5, no. 2 (1828), pp. 147–154.

6. F. Geirot, "O nervnoi goriachke (Febris nervosa) et Typha," *Voenno-Meditsinskii Zhurnal* 1, no. 1 (1823), pp. 1–74.

7. Tchudnovskii, "Mediko-topograficheskoe opisanie Sarapul'skago uezda Viatskoi gubernii," *Zhurnal Ministerstva Vnutrennykh Del*, April 1849, pp. 174–196.

8. Anon., "Mediko-topograficheskiia svedeniia o Kazanskoi gubernii," *Zhurnal Ministerstva Vnutrennykh Del* 14, no. 10 (1834), pp. 36–37; G. M. Sidorenko-Zelezinskaia, "O pervykh mediko-topograficheskikh opisaniia Kievskoi gubernii," *Sovetskoe Zdravookhranenii* 1965, no. 4 (1965), pp. 66–70; A. I. Drzhevetskii, "Mediko-topografiia Ust' Sysol'skago uezda Vologodskoi gubernii," *Mediko-Topograficheskii Sbornik*, ed. G. Arkhangel'skii (St. Petersburg, 1871), vol. 2, pp. 540–550.

9. A. Vilesov, "Klinicheskaia danniia iz nabliudenii zloi korchi, byvshei v 1896 g. v Solikamskom uezde, Permskoi gubernii," *Feld'sher* 6 (1896), p. 586.

10. G. A. Kolosov, "Ob epidemiiakh zloi korchi (ergotizma) i bor'be s nimi v prezhnee vremia i v poslednie gody," *Russkii Vrach* 10 (1911), pp. 1803–1807, 1842–1847.

11. Z. M. Agranovskii, "Voprosy gigieny pitaniia v russkikh rukopisnykh lechebnikakh XVII i XVIII vv.," *Trudy Leningradskogo sanitarno-gigienicheskogo meditsinskogo instituta* 14 (1953), pp. 211–219.

12. L. L. Klaf et al., *Ergotizm* (Kharkov, 1933), p. 9; Kolosov, "Ob epidemiiakh," *Russkii Vrach* 10 (1911), p. 1842.

13. Z. Zinin et al., *Izsledovaniia o sporyn'e (Secale cornutum) o sposobakh otkrytiia v muke* (St. Petersburg, 1864); A. Grunfeld, "Ob ergotizme v Rossii," *Obshchestvenno-sanitarnoe Obozrenie* 1 (1896), pp. 386–388 and 409–411; Kolosov, "Ob epidemiiakh," *Russkii Vrach* 10 (1911), pp. 1803–1807, 1842–1847, 11 (1912), pp. 55, 85, 120, 161, 198, and 236, and 12 (1913), pp. 182 and 221.

14. V. S. Lechnovich, "K istorii kul'turi kartofelia v Rossii," *Materialy po istorii zemledelia SSSR* 2 (1956), pp. 358–400.

15. N. N. Ulashchik, "Izmeneiia v khoziaistve krepostnoi Litvy i Zapadnoi Belorussii v sviazi s vvedeniam novykh kul'tur (kartofel')," *Materialy po istorii sel'skogo khoziaistva i krest'ianstva SSSR* 5 (1962), pp. 308–337.

16. E. S. Karnaukova, *Razmeshchenie sel'skogo khoziaistva Rossii v periode kapitalizma (1860–1914 gg.)* (Moscow, 1951), pp. 197–199.

17. S. A. Klepikov, *Pitanie russkogo krest'ianstva* (Moscow, 1920). See also Basile Kerblay, "L'évolution de l'alimentation rurale en Russie, 1896–1960," in *Pour une histoire de l'alimentation*, ed. J. J. Hemardinquer (Paris, 1970), pp. 188–211.

18. Roger A. Clarke and D. J. I. Matko, *Soviet economics facts, 1917–81* (New York, 1983).

19. Dana G. Dalrymple, "The Soviet famine of 1932–1934," *Soviet Studies* 15 (January 1964), pp. 250–284.

20. S. V. Vladimirskii, "Geograficheskoe rasprostranenie i zony vredonosnogo znachenie sporyn'i na rzhi v SSSR," *Sovetskaia Botanika* 5 (1939), pp. 77–87.

21. N. I. Ostrovskii et al., "Resursy sporyn'i v SSSR," *Rastitel'nye Resursy* 4, no. 2 (1968), pp. 162–172, and no. 4 (1968), pp. 468–477.

22. A. J. Coale, B. A. Anderson and E. Harm, *Human fertility in Russia since the nineteenth century* (Princeton, 1979); B. A. Anderson, *Internal migration during modernization in late nineteenth century Russia* (Princeton, 1980); L. S. Kaminskii, *Meditsinskaia i Demograficheskaia Statistika* (Moscow, 1974), p. 102.

23. B. R. Mitchell, *European historical statistics*, 2d ed. (New York, 1980), pp. 138–139.

24. Russia, Tsentral'nyi Statisticheskii Komitet, *Obshchii Svod po Imperii Resultatov Razrabotki Dannykh Pervoi Vseobshchei Perepisi Naseleniia (28 ianvaria 1897 goda)*, vol. 1 (St. Petersburg, 1905), p. iii.

25. Peter Czap, Jr., " 'A large family: The peasant's wealth': Serf households in Mishino, Russia, 1814–1858," in *Family formation in historic Europe*, ed. Richard Wall (Cambridge, England, 1983), pp. 105–150; Steven L. Hoch, "Serfs in Imperial Russia: Demographic insights," *Journal of Interdisciplinary History* 13, no. 2 (1982), pp. 221–246; John Hajnal, "European marriage patterns in perspective," in *Population in history*, ed. D. V. Glass and D. E. C. Eversley (Chicago, 1965), pp. 101–143; Hajnal, "Two kinds of pre-industrial household formation systems," in *Family forms in historic Europe*, ed. Richard Wall (Cambridge, 1983), pp. 65–104.

26. V. I. Grebenshchikov, "Plodovitost' zhenshchin v 26 guberniiakh Evropeiskoi Rossii po sravneniia s plodovitost'iu v zapadno-evropeiskikh gosudarstvakh," *Vestnik Obshchestvennoi Gigieny* 1904, pp. 1283–1303 and 1449–1463.

27. Steven L. Hoch, "Serf diet in nineteenth century Russia," *Agricultural History* 56, no. 2 (1982), pp. 391–414; D. J. Oddy, "Working class diets in late nineteenth century Britain," *Economic History Review* 23, no. 2 (1970), pp. 314–323.

28. Russia, Tsentral'noe Statisticheskoe Upravlenie (hereafter TSU), *Statistiki po Rossiiskoi Imperii; Dvizhenie Naseleniia v Evropeiskoi Rossii za 1896 g.* (St. Petersburg, 1899); TSU, *Dvizhenie Naseleniia v Evropeiskoi Rossii za 1897 g.* (St. Petersburg, 1900); TSU, *Dvizhenie Naselenniia v Evropeiskoi Rossii za 1898 g.* (St. Petersburg, 1903); TSU, *Obshchii Svod po Iperii Resultatov Razrabotki Dannykh Pervoi Vseobshchei Perepisi Naseleniia (18 ianvaria 1897 goda)* (St. Petersburg, 1905); A. G. Rashin, *Naselenie Rossii za 100 Let* (Moscow, 1956), pp. 165–166, 184–185, and 195–196. The 1897 census population totals were the base for estimating midyear population in 1896 and 1898.

29. Russia, Glavnaia Fizicheskaia Observatoriia (hereafter GFO), *Letopisi [Annalen]* (St. Petersburg, 1882–1898); GFO, *Klimatologicheskii Spravochnik po SSSR*, vol. 1 (*Evropeiskaia Chast' SSSR*) (Leningrad, 1932).

30. Russia, Tsentral'nyi Statisticheskii Komitet (hereafter TSK), *Résultats généraux de la recolte en Russie, 1890–1903* (St. Petersburg, 1909); TSK, *Résultats généraux de la récolte en Russie en 1909* (St. Petersburg, 1910).

31. Russia, Glavnoe Upravlenie Zemleustroistva i Zemledeliia, *Svod Statisticheskiikh Svedenii po Sel'skomy Khoziaistvu Rossii k Kontsu XIX Veka*, vol. 3 (St. Petersburg, 1906), pp. 89–128; A. Arnol'dov, "Ob izmeneniiu rzhanoi muki pri roste v nei nekotorykh plesnevykh gribov," *Vestnik Obshchestvennoi Gigieny*, October, 1905, pp. 1499–1521.

32. S. Glebovskii, "Satistika (po perepisi 1897 g.) etnografii v voprose o smertnosti v Rossii detei do odnogo goda," *Vestnik Obshchestvennoi Gigiena* 199 (1904), pp. 186–208.

33. Russia, Tsentral'nyi Statisticheskii Komitet, *Dvizhenie Naseleniia v Evropeiskoi Rossii za 1896 g. (. . . 1897 g., . . . 1898 g.)* (St. Petersburg, 1899, 1900, 1903).

34. George Barger, *Ergot and ergotism* (London, 1931), p. 39.

35. Anderson, *Internal migration*, pp. 200–201. The migration index equals the difference between male and female migration to Moscow plus the difference between male and female migration to St. Petersburg.

36. Glebovskii, "Statistika," pp. 193–194 and 201. Included were "Turco-Tatars" in Astrakhan, Kazan, Ufa, Orenburg, Penza, and Saratov provinces who were described as having distinctive customs of infant feeding.

37. A. A. Borisov, *Climates of the U.S.S.R.* (Chicago, 1965), p. 36.

38. Coale, Anderson, and Harm, *Human fertility in Russia*, pp. 22–23.

39. A. J. Russell (League of Nations), *A Memorandum on the epidemiology of cholera* (Geneva, 1925), p. 25.

40. B. V. Karaln'nik, N. M. Nurkina, and T. E. Kruglaia, "Analiz sezonnosti ostrykh kishechnykh zabolevanii s uchetom vyiavleniia dizenterii po indikatsii antigena," *Zhurnal Mikrobiologii, Epidemiologii, i Immunobiologii* 8 (1984), pp. 37–40.

41. Mitchell, *European historical statistics*, pp. 114–119.

42. H. H. Lamb, *Climate: past, present and future*, vol. 2 (London, 1977), p. 594.

43. See chapter 3 and A. R. Spoof, *Om Forgiftningar med Secale cornutum* (Helsinki, 1872).

44. M. A. Sarkisova, S. S. Shain, and L. I. Britvenko, "Poisk novykh shtammov sporyn'i—produtsentov peptidnykh ergoalkaloidov," *Mikologiia i Fitopatologiia* 17, no. 3 (1983), pp. 202–205.

45. B. Kosir, P. Smole, and Z. Povsic, ["Factors affecting the yield and quantity of sclerotia from Claviceps purpurea"], *Farmatsevtski Vestnik* 32 (1981), pp. 21–25.

46. N. I. Ostrovskii et al., "Alkaloidnost' i rasprostraneni e sporyn'i v SSSR," *Aptechnoe Delo* 1 (1959)), pp. 29–34; Ostrovskii et al., "Polevaia kul'tura sporyn'ia v SSSR," *Meditsinskaia Promyshlennost' SSSR* 12 (1959), pp. 11–15; Ostrovskii et al., "Sporyn'ia dikorastushchikh zlakov i gibridnykh rastenii kak vozmozhnyi material dlia selektsii shtammov Claviceps purpurea Tulasne," *Meditsinskaia Promyshlennost' SSSR* 9 (1964), pp. 46–48; Ostrovskii et al., "Resursy sporyn'i v SSSR," *Rastitel'nye Resursy* 4, no. 2 (1968), pp. 162–172, and 4, no. 4 (1968), pp. 468–477.

47. E. S. Zabolotnaia, "Soderzhanie alkaloidov v dikorastushchei sporyn'e v zavisimosti ot raionov proizrastaniia," *Trudy Vsesoiuznogo Nauchno-Issledovatel'skogo Instituta Lekarstvennykh i Aromaticheskikh Rastenii* 11 (1959), pp. 254–266; S. V. Vladimirskii, "Geograficheskoe rasprostranenie i zony vredonostnogo znacheniia sporyn'i na rzhi v SSSR," *Sovetskaia Botanika* 5 (1939), pp. 77–87.

48. I. A. Kurmanov, "Griby roda Fusarium po kormam nekotorykh zony SSSR," *Problemy Veterinarnoi Sanitarii* (Moscow, 1971), pp. 53–56.

3. A NEW LOOK IN THE DISTANT MIRROR

1. Carlo M. Cipolla, *Fighting the plague in seventeenth century Italy* (Madison, 1981), p. 100; V. S. Klimenko, "O chumnykh epidemiiakh v Rossii," *Voenno-Meditsinskii Zhurnal* 229 (September–December 1910), pp. 659–663.

2. Graham Twigg, *The Black Death: A biological appraisal* (London, 1984), pp. 60, 192, and 202.

3. C. J. Isselbacher, ed., *Harrison's principles of internal medicine*, 9th ed. (New York, 1980), pp. 662–665; Robert Pollitzer, *Plague* (Geneva, 1954).

4. Mary F. and T. H. Hollingsworth, "Plague mortality rate by age and sex in the parish of St. Botolph's without Bishopsgate, London, 1603," *Population Studies* 25, no. 1 (1971), pp. 131–146.

5. Leslie Bradley, "The most famous of all English plagues: A detailed analysis of the plague at Eyam, 1665–6," in *The plague reconsidered* (Matlock, 1977), pp. 63–94.

6. David Herlihy and Christiane Klapisch-Zuber, *Les Toscans et leur familles* (Paris, 1978), p. 463.

7. J. M. Mann et al., "Pediatric plague," *Pediatrics* 69, no. 6 (1982), pp. 762–767.

8. J. F. D. Shrewsbury, *A history of the bubonic plague in the British Isles* (Cambridge, 1970); R. Pollitzer and K. F. Meyer, "The ecology of plague," in *Studies in disease ecology*, ed. J. M. May (New York, 1961), p. 460; B. Bennasar and J. Goy, "Contribution à l'histoire de la consommation alimentaire du XIVe au XIXe siècle," *Annales* 30 (1975), pp. 402–409; Bridget Henisch, *Fast and feast: food in medieval society* (University Park, Pa., 1978); L. Stouff, *Ravitaillement et alimentation en Provence aux XIVe–XVe siècles* (Paris, 1970).

9. V. V. Rukhliada, "Biosintez T-2 toksina gribom Fusarium Sporotrichiella Bilai na razlichnykh rastitel'nykh substratakh," *Mikologiia i Fitopatologiia* 18 (1) (1984), pp. 73–75.

10. J. N. Biraben, *Les hommes et la peste en France et dans les pays européens et méditerranéens*, vol. 1 (Paris, 1975). This correlation was first observed by C. Martin, *Beiträge zur Chronologie und Ätiologie der Pest* (Weimar, 1879).

11. Biraben, *Les hommes et la peste*; Alfonso Corradi, *Annali delle Epidemie occorse in Italia*, vol. 1 (Bologna, 1865), p. 186; Pierre Alexandre, *Le climat au moyen âge en Belgique et dans les régions voisins* (Liège-Louvain, 1976); V. V. Betin, *Surovost' Zim v Evrope i Ledovitost' Baltiki* (Leningrad, 1962); C. E. Britton, *A meteorological chronology to A.D. 1450* (London, 1937); J. Z. Titow, "Evidence of weather in the account rolls of the Bishopric of Winchester, 1209–1350," *Economic History Review*, 2d ser., 12 (1960), pp. 360–407.

12. Ralph Josselin, *The Diary of Ralph Josselin, 1616–1683* (London, 1976), pp. 510–531.

13. William P. MacArthur, "The identification of some pestilences recorded in the Irish annals," *Irish Historical Studies* 6, no. 23 (1949), p. 179; L. M. Cullen, "Population growth and diet, 1600–1850," in *Irish population,*

economy and society, ed. J. M. Goldstrom and L. A. Clarkson (Oxford, 1981), pp. 89–112.

14. Twigg, *The Black Death*, pp. 60 and 111.

15. S. A. Barnett, *The rat: A study in behaviour* (Chicago, 1963).

16. Twigg, *The Black Death*, pp. 52–53 and 98.

17. E. Le Clainche, *Histoire de la médecine vétérinaire* (Toulouse, 1936), p. 124.

18. Johan Huizinga, *The waning of the Middle Ages* (New York, 1954), pp. 33–34.

19. H. Lamb, *Climate: past, present, and future*, vol. 2 (London, 1977), pp. 455–457.

20. Walter Janssen, *Studien zur Wüstungsfrage im fränkischen Altsiedelland zwischen Rhein, Mosel und Eifelnordrand* (Cologne, 1975), vol. 2, pp. 198–201.

21. Guy Bois, *Crise de féodalisme* (Paris, 1976), pp. 53–72; Leopold Delisle, *Etudes sur la condition de la classe agricole et l'état de l'agriculture en Normandie* (Evreux, 1851), p. 189.

22. Arlette Higounet-Nadal, *Périgueux aux XIVᵉ et XVᵉ siècles* (Bordeaux, 1978), pp. 143–160.

23. Herlihy and Klapisch-Zuber, *Les Toscans*, pp. 170–185 and 191.

24. Robert S. Gottfried, *Epidemic disease in fifteenth century England* (New Brunswick, N.J., 1978), pp. 127–138.

25. Shrewsbury, *A history of the bubonic plague*, p. 144.

26. Gottfried, *Epidemic disease*, pp. 99–117.

27. Marie-Thérèse Lorcin, *Les campagnes de la région lyonnaise aux XIVᵉ et XVᵉ siècles* (London, 1974), pp. 434, 526, and 529.

28. Charles Creighton, *History of epidemics in Britain* (New York, 1965), vol. 1, pp. 59–62; Justus Hecker, *The epidemics of the Middle Ages* (London, 1859), pp. 58–64, 174–175, and 353–360; George L. Kittredge, *Witchcraft in Old and New England* (New York, 1929), pp. 125–126; Gottfried, *Epidemic disease*, p. 44. For data on the weather: Titow, "Evidence of weather in the account rolls of the Bishopric of Winchester"; J. Z. Titow, "Le climat à travers les rôles de compatabilité de l'Evêche de Winchester, 1350–1450," *Annales* 25 (1970), pp. 312–350; Alexandre, *Le climat au moyen âge*; V. V. Betin, *Surovost' Zim v Evrope i Ledovitost' Baltiki*; Britton, *A meterological chronology*.

29. Richard Kieckhefer, *European witch trials* (Berkeley, 1976), pp. 106–147.

30. Lamb, *Climate*, pp. 564 and 593.

4. MYCOTOXINS AND HEALTH IN EARLY MODERN EUROPE

1. Thomas Willis, *The London practice of physick* (London, 1685), pp. 273–278.

2. Thomas Sydenham, *The entire works of Dr. Thomas Sydenham* (London, 1769), pp. 287–291.

3. Thomas Short, *A general chronological history of the air, weather,*

seasons, meteors, etc. in sundry places and different times (London, 1749), vol. 2, p. 441; James Johnstone, *A historical dissertation concerning the malignant epidemical fever of 1756* (London, 1758), pp. 17–18, and *A treatise on the malignant angina or putrid and ulcerous sore throat* (Worcester, 1779), pp. 61–64.

4. John Huxham, *Dissertation on the malignant ulcerous sore throat* (London, 1757); William Hillary, *Essay on smallpox* (London, 1740); John Fothergill, *An account of the sore throat attended with ulcers* (London, 1748).

5. Justus Hecker, *The epidemics of the Middle Ages* (London, 1844), pp. 201–202.

6. Short, *A general chronological history*, vol. 1, p. 204.

7. V. Perez Moreda, *Las crisis de mortalidad en la España interior (siglos XVI–XIX)* (Madrid, 1980); Joaquin de Villalba, *Epidemiologia Española*, 2 vols. (Madrid, 1802); Short, *A general chronological history*; Thomas Short, *A comparative history of the increase and decrease of mankind in England and several countries abroad* (London, 1767); Noah Webster, *A brief history of epidemic and pestilential diseases* (Hartford, 1799).

8. Thomas Cogan, *The haven of health* (London, 1636), pp. 28–29.

9. Fynes Moryson, *An itinerary*, vol. 4: 1617 (Glasgow, 1908), p. 171.

10. William Harrison, *The description of England* (1587; Ithaca, 1968), p. 133; Eric Kerridge, *The farmers of Old England* (Totowa, N.J., 1973), p. 163.

11. Tobias Venner, *Via recta ad vitam longam* (London, 1620), p. 19.

12. Anon., *Directions for health, naturall and artificial* (London, 1626), p. 18.

13. Ralph Josselin, *The diary of Ralph Josselin, 1616–1683* (London, 1976), p. 46; Joan Thirsk, ed., *The agrarian history of England and Wales*, vol. 4: 1500–1640 (Cambridge, 1967), p. 46.

14. J. K. Fedorowicz, *England's Baltic trade in the early seventeenth century* (Cambridge, 1980), pp. 111–114.

15. N. S. B. Gras, *The evolution of the English corn market from the twelfth to the eighteenth century* (Cambridge, Mass., 1915), pp. 101–103; W. S. Unger, "Trade through the Sound in the seventeenth and eighteenth centuries," *English Historical Review*, 2d ser., 12 (1959), pp. 206–221; J. A. Faber, "Decline of the Baltic grain trade in the second half of the seventeenth century," *Acta Historica Nederlandica* 1 (1966), pp. 108–131.

16. D. C. Coleman, *The economy of England, 1450–1750* (London, 1977), pp. 113–125; William Ashley, "The place of rye in the history of English food," *Economic Journal* 31 (1921), pp. 285–308; J. A. Yelling, "Changes in crop production in East Worcestershire, 1540–1867," *Agricultural History Review* 21 (1973), pp. 18–34.

17. G. Wood and J. R. Coley-Smith, "Observations on the prevalence and incidence of ergot disease in Great Britain, with special reference to open-flowering and male-sterile cereals," *Annals of Applied Biology* 95, no. 1 (1980), pp. 41–46.

18. Simon-André Tissot, "An account of the disease called ergot, in French, from its supposed cause, viz, vitiated rye," *Philosophical Transactions* 1 (1765), pp. 106–126.

19. Johann Taube, *Die Geschichte der Kriebelkrankheit* (Göttingen, 1783).

20. Charles Creighton, *History of epidemics in Britain* (New York, 1965), 2 vols.

21. Edward Jorden, *A disease called the suffocation of the mother* (London, 1603); Creighton, *History of epidemics*, vol. 1, pp. 505 and 572–574.

22. Robert Pemell, *A treatise on the diseases of children* (London, 1653), p. 6.

23. Walter Harris, *An exact inquiry* (London, 1693), p. 6.

24. Henry Newcome, *The compleat mother* (London, 1695), pp. 71–72.

25. Sydenham, *The entire works*.

26. Christopher C. Wilson, "Natural fertility in pre-industrial England, 1600–1799," *Population Studies* 38 (1984), pp. 225–240.

27. James E. Thorold Rogers, *A history of agriculture and prices in England (1259–1793)* (Oxford, 1866–1902), vol. 5, pp. 268–275.

28. Gordon Manley, "Central England temperatures: monthly means, 1659 to 1975," *Quarterly Journal of the Royal Meteorological Society* 100 (1974), pp. 389–405.

29. H. M. Van den Dool et al., "Average winter temperatures at de Bilt (the Netherlands), 1634–1977," *Climatic Change* 1 (1978), pp. 319–330.

30. H. H. Lamb, *Climate: Present, past and future*, vol. 2 (London, 1977); John H. Brazell, *London weather* (London, 1968); D. Justin Schove, "Summer temperature and tree rings in Scandinavia, A.D. 1461–1950," *Geografiska Annaler* 36, nos. 1–2 (1954), pp. 40–80.

31. E. A. Wrigley and R. S. Schofield, *The population history of England, 1541–1871* (Cambridge, Mass., 1981), pp. 28, 531–533.

32. B. R. Mitchell, *Abstract of British population statistics* (Cambridge, 1962), pp. 484–487 and 497–498; Wrigley and Schofield, *Population history of England*, pp. 642–644.

33. Willis, *London practice of physick*, pp. 137–150.

34. Antonia Fraser, *Cromwell, the Lord Protector* (New York, 1973).

35. John D. Post, "Climatic variability and the European Mortality wave of the early 1740's," *Journal of Interdisciplinary History* 15, no. 1 (1984), pp. 1–30.

5. WITCH PERSECUTION IN EARLY MODERN EUROPE

1. George L. Kittredge, *Witchcraft in Old and New England* (Cambridge, Mass., 1929), p. 163; C. L. Ewen, *Witchcraft and demonianism* (London, 1933), 93–98; *Sadducismus Defellatus* (London, 1698), pp. 2, 9, 19, 32, 38, and 41; and J. M. MacPherson, *Primitive beliefs in the northeast of Scotland* (New York, 1977), pp. 174–176, 181, and 187.

2. S. Broekhuizen, ed., *Atlas of the cereal growing areas of Europe*, vol. 2 (Wagenigen-Amsterdam, 1969).

3. M. L. Parry and T. L. Slater, eds., *The making of the Scotch countryside* (Montreal, 1980), p. 77; Robert Chamber, *Domestic annals of Scot-*

land, 2d ed., vol. 1 (London, 1859), pp. 266 and 290; Ian Whyte, *Agriculture and society in seventeenth century Scotland* (Edinburgh, 1979), p. 63.

4. Patrick F. Byrne, *Witchcraft in Ireland* (Cork, 1967); L. M. Cullen, "Population growth and diet, 1600–1850," in *Irish population, economy and society*, ed. J. M. Goldstrom and L. A. Clarkson (Oxford, 1081), pp. 89–112.

5. Carmela Lison-Tolosana, *Brujería, estructura social, y simbolismo en Galicia* (Madrid, 1979), pp. 108–109; Gustav Hennigson, *The witches advocate: Basque witchcraft and the Spanish Inquisition, 1609–1614* (Reno, Nevada, 1980); Owsei Temkin, *The falling sickness*, 2d ed. (Baltimore, 1971), p. 37.

6. E. William Monter, *Witchcraft in France and Switzerland* (Ithaca, 1976), p. 82; Gerhard Schormann, *Hexenprozesse in Nordwestdeutschland* (Hildesheim, 1977), p. 157.

7. H. Erik Midelfort, *Witch-hunting in southwestern Germany, 1562–1654* (Stanford, 1972); Wilhelm Abel, *Agricultural fluctuations in Europe* (New York, 1980); Fritz H. Schweingruber et al., "Dendroclimatic studies in conifers from central Europe and Britain," *Boreas* 8 (1979), pp. 427–452.

8. Zoltán Kovács, "Die Hexen in Russland," *Acta Etnographica Academiae Scientiarium Hungaricae* 22, nos. 1–2 (1973), pp. 51–87; R. Zguta, "Was there a witch craze in Muscovite Russia?" *Southern Folklore Quarterly* 41 (1977), pp. 119–128, and "Witchcraft trials in seventeenth century Russia," *American Historical Review* 82, no. 5 (1977), pp. 1187–1207.

9. Marie-Sylvie Dupont-Bouchet et al., *Prophètes et sorcières dans les Pays-Bas, XVIᵉ–XVIIIᵉ siècles* (Paris, 1978), p. 18.

10. August Hirsch, *Handbook of geographical and historical pathology* (London, 1883–86).

11. C. L. Ewen, *Witch hunting and witch trials, 1559–1736* (London, 1929), and *Witchcraft and demonianism*.

12. Alan MacFarlane, *Witchcraft in Tudor and Stuart England* (New York, 1970), esp. pp. 178–185.

13. Eric Kerridge, *The agricultural revolution* (London, 1967), pp. 223–227.

14. MacFarlane, *Witchcraft in Tudor and Stuart England*, pp. 181–183 and 304–309.

15. Ewen, *Witchcraft and demonianism*, pp. 93–94; Michael MacDonald, *Mystical Bedlam* (Cambridge, 1981), pp. 208–209.

16. Thomas Wright, *Narratives of sorcery and magic* (Detroit, 1971), pp. 254–276.

17. Ewen, *Witchcraft and demonianism*, pp. 191–192.

18. Edward Fairfax, *Demonologia* (London, 1622). Rye was an important crop in Norfolk as well as Essex at this time.

19. Reginald Scot, *The discoverie of witchcraft* (1584; Totowa, 1973), p. 6.

20. MacFarlane, *Witchcraft in Tudor and Stuart England*, pp. 147, 162–163, and 180.

21. Ibid., pp. 152–157.

22. William Harrison, *The description of England* (1587; Ithaca, 1968), p. 133.

23. Abbot L. Cummings, *Rural household inventories, 1675–1775* (Boston, 1964), pp. 4–72.

24. MacFarlane, *Witchcraft in Tudor and Stuart England*, pp. 8, 30, and 154–155; Kerridge, *Agricultural revolution*, pp. 59, 69, 71, 72, 77, 79, 83, 85, 137–138, and 176; William J. Ashley, *The bread of our forefathers* (Oxford, 1928), documents, indicating that rye was the predominant crop in the northeastern and southwestern corners of Essex, and map.

25. Harrison, *Description*, p. 133. See also Andrew B. Appleby, "Grain prices and subsistence crises in England and France, 1590–1740," *Journal of Economic History* 39 (1979), pp. 865–887.

26. Kerridge, *Agricultural revolution*, pp. 344–345, and *The farmers of Old England* (Totowa, N.J., 1973), p. 163. For crop changes elsewhere see Victor Skipp, *Crisis and development* (Cambridge, 1974), pp. 44–49.

27. Florence M. McNeill, *The Scots kitchen: Its lore and recipes*, 2d ed. (London and Glasgow, 1963), p. 178; Marjorie Plant, *The domestic life of Scotland* (Edinburgh, 1952), p. 100; and Whyte, *Agriculture and society*, p. 63.

28. Maud Grieve, *A modern herbal* (New York, 1931), vol. 1, p. 70, vol. 2, pp. 555–556; Richard Le Strange, *A history of herbal plants* (London, 1977), pp. 70–71 and 260–262; and Maria Leach, *Funk and Wagnalls standard dictionary of folklore, mythology and legend* (New York, 1950), pp. 732–733.

29. Keith Thomas, *Religion and the decline of magic* (New York, 1971), p. 548.

30. C. K. Sharpe, *A historical account of belief in witchcraft in Scotland* (1884; Detroit, 1974), p. 101.

31. H. H. Lamb, *Climate: Present, past, and future* (London, 1977), vol. 2, pp. 568–569.

6. THE GREAT FEAR OF 1789

1. Georges Lefebvre, *The Great Fear of 1789* (New York, 1973), pp. 147 and 156–169.

2. William Doyle, *Origins of the French Revolution* (Oxford, 1980), pp. 202–203; Albert Soboul, *The French Revolution, 1787–1799* (New York, 1975), pp. 145–150.

3. Soboul, *French Revolution*, pp. 144–147; Philippe-Jean Hesse, "Géographie coutumière et révoltes paysannes en 1789," *Annales historiques de la Révolution Française*, 1979, pp. 283–286.

4. Louis Jacob, "La Grande Peur en Artois," *Annales de la Révolution Française* 13 (1936), pp. 123–148; and Georges Bussière, *Etudes historiques sur la Révolution en Périgord*, part 3 (Paris, 1903), pp. 94–95.

5. Dr. Geoffrey, "Constitution de l'année 1789, avec le détail des maladies qui ont régné pendant les différents saisons de cette année," *Histoire et Mémoires de la Société Royale de Médecine*, 1789 (Paris, 1792), pp. 1–18.

6. L. E. Gachet and Dr. Maison, *Tableau historiques des événemens présens* (Paris, 1789), passim.

7. J. P. Goubert, *Malades et médecins en Bretagne, 1770–1790* (Paris, 1974), p. 217.

8. Jean Taxil, *Tracte de l'épilepsie* (Lyon, 1602), p. 253; Jacques Guillemeau, *De la nourriture et gouvernement des enfants* (Paris, 1609); and M. Baumes, *Des convulsions dans l'enfance* (Nîmes, 1789), pp. 89 and 313.

9. Dr. Baraillon, "Observations sur une espèce d'épilepsie qui reconnoit pour cause le virus exanthematique miliare," *Histoire et Mémoires de la Société Royale de Médecine* 1 (1776), pp. 225–231; and Dominique Villars, *Observations de médecine sur une fièvre épidémique qui a régné dans le Champsaur et le Valgaudemar en Dauphiné pendant les années 1779 et 1780* (Grenoble, 1781).

10. Joseph Raulin, *Traité des affections vaporeuses du sexe* (Paris, 1758), pp. 1–34.

11. Salerne, "Mémoire sur les maladies que cause le seigle ergoté," *Mémoires de Mathématique et Physique* 2 (1755), pp. 155–163; M. N. P. Vetillart du Ribert, *Mémoire sur une espèce de poison* (Paris, 1770); Al.-Phil. Read, *Traité du seigle ergoté* (Strasbourg, 1771); Saillant, "Recherches sur la maladie convulsive épidémique," *Histoire et Mémoires de la Société Royale de Médecine* 1 (1776), pp. 303–320; Le Brun, "Sur l'effet des seigles de mauvaise qualité," *Histoire et Mémoires de la Société Royale de Médecine* 2 (1777–1778), pp. 299–302; and Abbé A. H. Tessier, *Traité des maladies des grains* (Paris, 1783). For more recent studies of ergot in eighteenth-century France, see André J. Bourde, *Agronome et agronomes en France au XVIIIᵉ siècle*, vol. 2 (Paris, 1967), pp. 599 and 611–616; and Christian Poitou, "Ergotisme, ergot de seigle, et épidémies en Sologne au XVIIIᵉ siècle," *Revue d'Histoire Moderne et Contemporaine* 23 (1976), pp. 354–368.

12. Johann Taube, *Die Geschichte der Kriebelkrankheit* (Göttingen, 1783), reviewed in *Journal de Médecine* 44 (1785), pp. 288–300.

13. N. N. Reformatskii, *Dushevnoe Rasstroitstvo pri Otravlenii Sporyn'ei (Bolezn' 'zlaia korcha')* (Moscow, 1893), pp. 365–368.

14. Jean-Claude Toutain, *La consummation alimentaire en France de 1789 à 1964* (Geneva, 1971), p. 2031; F. Mayer, "Evolution de l'alimentation des français, 1781–1972," *Gastroentérologie Clinique et Biologique* 1 (1977), pp. 1043–1051; and Hugh Clout, *Agriculture in France on the eve of the Railway Age* (London, 1980), p. 208.

15. On climate, see Geoffrey, "Constitution de l'année 1789," pp. 5–6; and *Journal de Médecine* 79 (1789), p. 428. For the ripening times of rye in various parts of France, see Comité Nationale de Géographie, *Atlas de France* (Paris, 1933–1945), map. 13. For an overview of winter temperature and spring rainfall, 1697–1789, see Gordon Manley, "Central England temperatures: monthly means, 1659 to 1973," *Quarterly Journal of the Royal Meteorological Society* 100 (1974), pp. 389–405; and B. G. Wales-Smith, "An analysis of monthly rainfall totals, 1697–1970," *Meterological Magazine* 100 (1971), pp. 345–362.

16. For distribution of panic symptoms, see Lefebvre, *Great Fear of 1789*, p. 4; for distribution of crops, see Arthur Young, *Travels in France during*

the years 1787, 1788 and 1789 (Cambridge, 1950); Abbé Tessier, "Mémoire sur les substances farineuses dont on fait du pain dans les diverses parties de la France," *Histoire et Mémoires de la Société Royale de Médecine* 10 (1789), pp. xci–clv; and Clout, *Agriculture in France*, p. 100.

Maslin normally contained two-thirds wheat, one-third rye. The amount of ergot by weight in unharvested grain is 43 percent as great as the number of ergotized plants in the field. If one-twelfth (8.3 percent) of all ears of rye were ergotized in 1789, then the unharvested maslin might contain 1.2 percent ergot, more than sufficient to produce symptoms, and four times the amount currently considered dangerous to humans in the United States. The map was constructed on the basis of production of rye and maslin in 1836–1838, on the assumption that this distribution was closer to the pattern of consumption in 1789, when the market was less developed, than consumption data for 1836–1838.

17. T. J. A. Le Goff, *Vannes and its region* (Oxford, 1981), pp. 5–17; and Donald Sutherland, *The Chouans* (Oxford, 1982), pp. 66 and 146–147.

18. Young, *Travels*, p. 281.

19. Gerard Bouchard, *Le village immobile* (Paris, 1972), p. 112.

20. Albert Soboul, *Problèmes paysans de la révolution, 1789–1848* (Paris, 1976), p. 272.

21. Louis Lepecq de la Cloture, *Collection d'observations sur les maladies, . . .* (Rouen and Paris, 1778), pp. 29, 200, 208, 273, 292, 315, 356, 462, 468, 756, and 921; and J. A. F. Ozanam, *Histoire médicale*, 2d ed., vol. 4 (Paris, 1835), p. 217.

22. Clarke Garrett, *Respectable folly; millennarians and the French Revolution in France and England* (Baltimore, 1975), pp. 17, 26, 40, 111, and 160, Robert Whytt, *Observations on the nature, cause and cure of those disorders which have been commonly called nervous, hypochondriac, or hysteric* (Edinburgh, 1767), pp. 212–213; Hugo Froderberg, "The preaching sickness in Smaland, Sweden in the 1840s," *Sydsvenska medicinhistoriska Sallskapet Arsskrift* (1965), pp. 79–94; E. L. Backman, *Religious dances in the Christian Church and in popular medicine* (London, 1952), pp. 310–311; and Mary K. Matossian, "Religious revivals and ergotism in America," *Clio Medica* 16, no. 4 (1982), pp. 185–192.

23. L. Del Panta and M. Livi Bacci, "Chronologie, intensité, et diffusion des crises de mortalité en Italie, 1600–1850," *Population*, num. spéc. 32 (1977), pp. 401–446.

24. John Ferriar, *Medical histories and reflections* (London, 1810–1813), vol. 1, pp. 153–156; vol. 2, p. 57.

25. John F. C. Harrison, *The Second Coming* (New Brunswick, N.J., 1979), p. 218; Garrett, *Respectable folly*, pp. 111 and 160.

26. T. M. Wigley et al., "Spatial patterns in precipitation in England and Wales," *Journal of Climatology* (forthcoming).

27. George Barger, *Ergot and ergotism* (London, 1931), pp. 73–77.

7. THE POPULATION EXPLOSION OF 1750–1850

1. Colin McEvedy and Richard Jones, *Atlas of world population history* (New York, 1978), p. 28.

2. E. A. Wrigley and R. S. Schofield, *The population history of England, 1541–1871* (Cambridge, Mass., 1981); E. A. Wrigley, "The growth of population in eighteenth-century England: a conundrum resolved," *Past and Present* 98 (1983), pp. 121–150; R. S. Schofield, "The impact of scarcity and plenty on population change in England, 1541–1871," *Journal of Interdisciplinary History* 14 (1983), pp. 265–291.

3. Peter H. Lindert, "English living standards, population growth, and Wrigley-Schofield," *Explorations in Economic History* 20, no. 2 (1983), pp. 131–155; David R. Weir, "Life under pressure: France and England, 1670–1870," *Journal of Economic History* 44 (March 1984), pp. 27–47; Myron F. Gutmann and Rene Leboute, "Rethinking protoindustrialization and the family," *Journal of Interdisciplinary History* 14 (1984), pp. 587–609; J. A. Goldstone, "The demographic revolution in England: a re-examination," *Population Studies* 40 (1986), pp. 5–33.

4. C. H. Pouthas, *La population française pendant la première moitié du XIXe siècle* (Paris, 1956), pp. 208–209; Jacques Houdaille, "Célibat et âge du mariage aux XVIIIe et XIXe siècles en France," *Population* (1978), pp. 43–84, and (1979), pp. 403–442.

5. Thomas McKeown, "Fertility, mortality, and causes of death," *Population Studies* 32 (1978), p. 537, and *The modern rise of population* (New York, 1976), pp. 34–40.

6. R. E. Jones, "Further evidence of the decline in infant mortality in pre-industrial England: North Shropshire, 1561–1810," *Population Studies* 34 (1980), pp. 239–250; D. E. C. Eversley, "A survey of the population in an area of Worcestershire from 1660 to 1850 on the basis of parish registers," *Population Studies* 10 (1957), pp. 253–279; E. A. Wrigley, "Mortality in pre-industrial England: the example of Colyton, Devon, over three centuries," in *Population and social change*, ed. D. V. Glass and D. E. C. Eversley (London, 1972), pp. 244–273; T. H. Hollingsworth, "The demography of the British peerage," *Population Studies*, supplement, 18, no. 2 (1964), pp. 56–57; Wrigley and Schofield, *Population history of England*, pp. 249–250.

7. Michael Drake, *Population and society in Norway, 1735–1865* (Cambridge, 1969), pp. 54–65, 95; Erland A. G. von Hofsten and H. Lundstrom, *Swedish population history: Main trends from 1750 to 1970* (Stockholm, 1976), pp. 15, 46–47; K. H. Connell, *The population of Ireland, 1750–1845* (London, 1975); Michael Drake, "The Irish demographic crisis of 1741," *Historical Studies* 6 (1968), pp. 101–126; Cormac O'Grada, "The population of Ireland, 1700–1900: A survey," *Annales de Démographie Historique* (1979), pp. 281–299.

8. James C. Riley, "Insects and European mortality decline," *American Historical Review* 91 (1986), pp. 833–858.

9. Susan Cott Watkins and E. van de Walle, "Nutrition, mortality and population size: Malthus' court of last resort," *Journal of Interdisciplinary History* 14 (1983), pp. 205–227; Thomas McKeown, *The modern rise of population* (New York, 1976); and Jean-Noël Biraben, "Alimentation et démographie historique," *Annales de Démographie Historique* (1976), pp. 22–40.

10. See especially F. Melier, "Etude sur les subsistences envisagées dans leurs rapports avec les maladies et la mortalité," *Mémoires de l'Académie*

Royale de Médecine 10 (1843), pp. 170–205; William L. Langer, "Europe's initial population explosion," *American Historical Review* 69 (1963), pp. 1–17; Langer, "American foods and Europe's population growth," *Journal of Social History* 8, no. 2 (1976), pp. 51–66; Redcliffe N. Salaman, *The history and social influence of the potato* (Cambridge, 1949).

11. H. H. Lamb, *Climate: Present, past, and future*, vol. 2 (London, 1977), pp. 593–595; Ronald Lee, "Short-term variation: vital rates, prices, and weather," in *The population history of England, 1541–1871*, by E. A. Wrigley and R. S. Schofield (Cambridge, Mass., 1981), pp. 356–401; Ronald Lee and Toni Richards, "Weather, nutrition, and the economy: short-run fluctuation in births, deaths, and marriage, France, 1740–1909," *Demography* 20 (May 1983), pp. 197–212.

12. Eric Kerridge, *The farmers of Old England* (Totowa, N.J., 1973), p. 163.

13. Charles Deering, *Nottingham Vetus et Nova* (Nottingham, 1751), p. 72.

14. E. J. T. Collins, "Dietary change and cereal consumption in Britain in the nineteenth century," *Agricultural History Review* 23, pt. 2 (1975), pp. 98–100; Capel Lofft, "Food of the poor of Ingelton," [Yorkshire] *Annals of Agriculture* 26 (1796), p. 226.

15. Salaman, *History and social influence of the potato*, pp. 518–532.

16. H. H. Lamb, "Volcanic dust in the atmosphere, with a chronology and assessment of its meteorologic significance," *Philosophical Transactions of the Royal Society of London*, ser. A, 266 (1970), pp. 425–533; Sigurdur Thorarinsson, "On the damage caused by volcanic eruptions with a special reference to tephra and gas," in *Volcanic activity and human ecology*, ed. P. H. Sheets and D. K. Grayson (New York, 1979), pp. 125–159; C. U. Hammer, H. B. Clausen, and W. Dansgaard, "Greenland ice sheet evidence of post-glacial volcanism and its climatic impact," *Nature* 288 (November 20, 1980), pp. 230–235; Wrigley and Schofield, *Population history of England*, p. 653; Michael Flinn, ed., *Scottish population history* (Cambridge, 1977), pp. 235–236 and 373–374; J. Mokyr, "Industrialization and poverty in Ireland and the Netherlands," *Journal of Interdisciplinary History* 10 (1980), pp. 429–458.

17. Hollingsworth, "Demography of the British peerage," p. 42; Wrigley and Schofield, *Population history of England*, p. 249.

18. Thomas Phayer, *The boke of children* (London, 1612), pp. 9 and 61; Robert Pemell, *A treatise on the disease of children* (London, 1653); Walter Harris, *A treatise on the acute diseases of infants* (London, 1689), and *An exact inquiry . . .* (London, 1693), p. 18; John Pechey, *A general treatise on the diseases of infants and children* (London, 1697); Henry Newcome, *The compleat mother* (London, 1695), p. 71.

19. George Cheyne, *The English malady, or a treatise on nervous diseases of all kinds* (London, 1733), pp. 163, 167, and 335–337; *The natural method of cureing the disease of the mind depending on the body* (London, 1742), pp. 90, 126, 128, 133, 185, 176–77; Anon., *The art of nursing* (London, 1733), p. 59.

20. Simon-André Tissot, "An account of the disease called ergot, in

French from its supposed cause, viz, vitiated rye," *Philosophical Transactions* 1 (1765), pp. 106–126.

21. William Grant, *Observations on the nature and cure of fevers*, 2d ed. (London, 1772), pp. 9 and 23.

22. William Moss, *An essay on the management and nursing of children* (London, 1781), pp. 73–74.

23. Randolph Trumbach, *The rise of the egalitarian family: Aristocratic kinship and domestic relations in eighteenth-century England* (New York, 1978), pp. 196–201.

24. Hollingsworth, "Demography of the British peerage," p. 56; Wrigley and Schofield, *Population history of England*, p. 230; T. A. Hollingsworth, "Mortality in British peerage families since 1600," *Population* 32 (1977), pp. 324–352.

25. P. H. Lindert and J. G. Williamson, "English workers' living standards during the Industrial Revolution," *Economic History Review* 36 (1983), pp. 1–25; Lindert and Williamson, "English workers' real wages: Reply to Crafts," *Journal of Economic History* 45 (1985), pp. 145–153; L. D. Schwarz, "The standard of living in the long run: London, 1700–1860," *Economic History Review* 38 (1985), pp. 24–41.

26. Charles Creighton, *A history of epidemics in England* (New York, 1965), vol. 2, pp. 3, 694, and 722–723; August Hirsch, *Handbook of geographic and historical pathology* (London, 1883–1886), vol. 3, p. 81.

27. Mary K. Matossian, "Death in London, 1750–1909," *Journal of Interdisciplinary History* 16 (Autumn 1985), pp. 183–197.

28. Jacques Dupaquier, *La population française aux XVIIe et XVIIIe siècles* (Paris, 1979), p. 82.

29. Jean-Claude Toutain, *La population de la France de 1700 à 1959* (Paris, 1963); Wrigley and Schofield, *Population history of England*, p. 214.

30. Dupaquier, *La population française*, p. 15.

31. B. G. Wales-Smith, "An analysis of monthly rainfall totals representative of Kew (Surrey), 1697–1970," *Meteorological Magazine* 100 (1971), pp. 357–58.

32. Gordon Manley, "Central England temperatures: Monthly means, 1659–1973," *Quarterly Journal of the Royal Meteorological Society* 100 (1974), pp. 389–405; Wales-Smith, "Analysis of monthly rainfall," pp. 345–62; M. E. Renou, "Etudes sur le climat de Paris," *Annales du Bureau Central Météorologique de France* 1 (1885, published in 1887), pp. 211–212.

33. Louis Lepecq de la Clôture, *Collection d'observations sur les maladies et constitution épidémiques* (Paris and Rouen, 1778), pp. 21 fn., 200, and 273.

34. Jean-Claude Perrot, *Genèse d'une ville moderne: Caen au XVIIIe siècle* (Paris, 1975), vol. 1, p. 146.

35. C. N. Lecat, "Mémoire sur les fièvres malignes qui régnèrent à Rouen à la fin de 1753 et au commencement de 1754," *Précis analytique des travaux de l'Académie des Sciences, Belles Lettres et Arts de Rouen, 1744–1803* (1816), vol. 2, pp. 77–81; Dr. Marteau, "Plusieurs maux de gorges gangreneux et épidémiques," *Journal de Médecine* 4 (1756), pp. 227–231; Dr. Bonté, "Description de la fièvre maligne épidémique qui a régné à Cou-

tances et dans ses environs pendant les années 1772 et 1773," *Mémoires de la Société Royale de Médecine* (1776), pp. 22–60; Dubosq de Laroberdière, "Recherches sur la scarlatina angineuse, contenant l'histoire de l'épidémie qui a régné à Vire dans les années VIII et IX (1800 et 1801)," *Journal Général de Médecine* 22 (1804–1805), pp. 430–443; Michel Asselin, *Examen analytique de la topographie et de la constitution médicale de l'arrondissement de Vire, Département de Calvados* (Caen, 1819), pp. 166 and 183; "Rapport général sur les épidemies qui ont regné en France depuis 1771 jusque'à 1830 exclusivement," *Mémoires de l'Académie Royale de Médecine* 3 (1833), p. 423.

36. Michel Morineau, *Les Faux-semblants d'un demarrage économique: agriculture et démographie en France au XVIII siècle* (Paris, 1971), pp. 294–296.

37. "Rapport général," pp. 377–429.

38. Dr. Boismare, "Mémoire sur la topographie et les constitutions médicales de la ville de Quillebeuf, et des lieux circonvoisins dont elle reçoit des influences," *Précis analytique des travaux de l'Académie des Sciences, Belles Lettres et Arts de Rouen* (1811), pp. 94–132, and (1812), pp. 65–107.

39. A. G. Ballin, "Renseignements statistiques sur la mortalité des enfants en bas âge," *Précis analytique . . . Rouen* 2 (1830), pp. 65–68.

40. Georges Duby and Armand Wallon, eds., *Histoire de France rurale*, vol. 3: Maurice Agulhon, Gabriel Désert, and Robert Speklin, *Apogée et crise de la civilisation paysanne* (Paris, 1976), pp. 108–112, 118; André Dubuc, "La culture de la pomme de terre en Normandie avant et depuis Parmentier," *Annales de Normandie* 3 (1953), pp. 50–68; Gabriel Désert, "La culture de la pomme de terre dans Calvados au XIX siècle," *Annales de Normandie* 5 (1955), pp. 251–270; Désert, *Les archives hospitalières: Sources d'histoire économique et sociale* (Caen, 1977), p. 158.

41. Christian Poitou, "La mortalité en Sologne orléanaise de 1670 à 1870," *Annales de Démographie Historique* (1978), pp. 235–264.

42. Manley, "Central England temperatures," pp. 395–396; Renou, "Etudes sur le climat de Paris," pp. 211–212.

43. Wrigley and Schofield, *Population history of England*, p. 253; and Dupaquier (1979), p. 62.

8. THE THROAT DISTEMPER

1. Ernest Caulfield, *A true history of the terrible epidemic vulgarly called the throat distemper* (New Haven, 1939), pp. 40–44.

2. William Douglass, *The practical history of a new epidemical eruptive military fever with an angina ulcusculosa* (Boston, 1736).

3. Kurt J. Isselbacher, *Harrison's principles of internal medicine*, 10th ed. (New York, 1983), pp. 930–931.

4. E. B. Smalley et al., "Mycotoxicoses associated with moldy corn," in *Toxic microorganisms, mycotoxins, botulism*, ed. Mendel Herzberg (Washington, D.C., 1970), pp. 163–175.

5. Douglass, *Practical history of a fever*; Jonathan Dickinson, *Obser-*

vations on that terrible disease vulgarly called the throat distemper (Boston, 1740).

6. Letter of William Douglass to Cadwallader Colden, November 12, 1739, in *The letters and papers of Cadwallader Colden*, 2 vols. (New York, 1918–1927), vol. 2, p. 197.

7. Regina Schoental, "Mouldy grain and the aetiology of pellagra: The role of toxic metabolites of *Fusarium*," *Biochemical Society Transactions* 8 (1980), pp. 147–150.

8. Joseph Forgacs and W. T. Carll, "Mycotoxicoses," *Advances in veterinary science* 7 (1962), p. 357.

9. Jabez Fitch, *An account of the numbers that have died of the distemper of the throat within the province of New Hampshire* (Boston, 1736).

10. Joshua Hempstead, *The diary of Joshua Hempstead* (New London, 1901), pp. 303–313.

11. William Willis, ed., *Journals of the Rev. Thomas Smith and the Rev. Samuel Deane* (Portland, Me., 1849), pp. 85–89.

12. Dickinson, *Observations on that terrible disease*, p. 1.

13. Samuel Lane, *A journal for the years 1739–1803* (Concord, N.H., 1937), p. 65.

14. Hempstead, *Diary*, pp. 282ff.

15. Robert R. Walcott, "Husbandry in colonial New England," *New England Quarterly* 9 (1936), p. 233.

16. Charles E. Clark, *The eastern frontier* (New York, 1970), p. 251.

17. John Kearsley, *Observations on the angina maligna, or, the putrid and ulcerous sore throat* (Philadelphia, 1769).

18. The Holyoke Papers, Essex Institute, Salem, Mass.

19. *Vital records of Andover, Massachusetts*, vol. 2 (Topsfield, Mass., 1912).

9. ERGOT AND THE SALEM WITCHCRAFT AFFAIR

1. L. R. Caporael, "Ergotism: The Satan loosed in Salem?" *Science* 192 (1976), pp. 21–26.

2. N. P. Spanos and J. Gottlieb, "Ergotism and the Salem Village witch trials," *Science* 194 (1976), pp. 1390–1394; N. P. Spanos, "Ergotism and the Salem witch panic," *Journal of the History of the Behavioral Sciences* 19 (October 1983), pp. 358–369.

3. P. Boyer and S. Nissenbaum, *Salem possessed: The social origins of witchcraft* (Cambridge, Mass., 1974).

4. John Demos, *Entertaining Salem* (Oxford, 1982).

5. Chadwick Hansen, *Witchcraft at Salem* (New York, 1969).

6. H. Mersky, *The analysis of hysteria* (London, 1979); F. Sirois, *Epidemic hysteria* (Copenhagen, 1974).

7. S. A. Green, "Salem witchcraft: A biopsychosocial analysis," *Pharos* 45, no. 3 (1982), pp. 9–13; and John T. Maltsberger, "Even in the twelfth generation—Huntington's chorea," *Journal of the History of Medicine and Allied Sciences* 16 (1961), pp. 1–17.

8. P. Boyer and S. Nissenbaum, eds., *The Salem witchcraft papers* (New

York, 1977), 3 vols.; Alan MacFarlane, *Witchcraft in Tudor and Stuart England* (London, 1970); C. L. Ewen, *Witchcraft and demonianism* (London, 1933); and R. Scot, *The discoverie of witchcraft* (Totowa, N.J., 1973).

9. N. N. Reformatskii, *Dushevnoe Rasstroistvo pri Otravlenii Sporyn'ei (Bolezn' Zlaia Korcha)* (Moscow, 1893), pp. 96, 102, 365–368.

10. Boyer and Nissenbaum, *Salem witchcraft papers*, vol. 1, p. 190.

11. Ibid., vol. 2, p. 595.

12. Ibid., vol. 2, pp. 601–602.

13. Hoffer and H. Osmond, *The hallucinogens* (New York, 1967); R. C. De Bold and R. C. Leaf, eds., *LSD, man and society* (Middletown, Conn., 1967); R. Blum, *Utopiates* (New York, 1964); and M. Tarshis, *The LSD controversy* (Springfield, Ill., 1972).

14. Samuel Wyllys, *Collection, Records of Trials for Witchcraft in Connecticut*, Connecticut State Library, Hartford; J. M. Taylor, *The witchcraft delusion in colonial Connecticut, 1647–1697* (Stratford, Conn., 1908).

15. G. A. Kolosov, "Geistesstörungen bei Ergotismus," *Archiv für Psychiatrie und Nervenkrankheit* 53 (1914), pp. 1118–1129, trans. B. Holmes and W. H. Sole in the *Chicago Medical Recorder* 38 (1916), p. 279.

16. T. Demeke, Y. Kidane, and E. Wuhib, "Ergotism: A report on an epidemic, 1977–78," *Ethiopian Medical Journal* 17 (1979), pp. 107–114.

17. *Danvers Historical Collections* (Danvers, Mass., 1926); "Diary of Josiah Green," *Essex Institute Historical Collections* 8 (1866), pp. 215–224; 10 (1869), pp. 73–104; and 36 (1900), pp. 323–330.

18. L. N. Eleutherius, "*Claviceps purpurea* on Spartina in coastal marshes," *Mycologia* 66 (1974), pp. 978–986; S. K. Harris, *The flora of Essex County, Massachusetts* (Salem, 1975).

19. Philip C. Lyons, Ronald D. Plattner, and Charles W. Bacon, "Occurrence of peptide and clavine ergot alkaloids in tall fescue grass," *Science* 232 (April 25, 1986), pp. 486–489.

20. Boyer and Nissenbaum, *Salem witchcraft papers*, vol. 2, pp. 410, 423, and 524.

21. G. L. Burr, ed., *Narratives of the witchcraft cases, 1648–1706* (New York, 1914); Cotton Mather, *Diary* (New York, n.d.).

22. Sarah F. McMahon, "A comfortable subsistence: The changing composition of diet in rural New England, 1620–1840," *William and Mary Quarterly* 62 (January 1985), p. 33.

23. Diary of Zaccheus Collins, Essex Institute, Salem, Mass.; A. L. Cummings, *Rural household inventories, 1675–1775* (Boston, 1964).

24. "Diary of Lawrence Hammond," *Massachusetts Historical Society Proceedings*, 2d ser., 7 (1891–1892), pp. 156–160.

25. Samuel Sewall, *Diary* (New York, 1972), vol. 1, p. 362.

26. Noah Webster, *A brief history of epidemics and pestilential diseases* (Hartford, Conn., 1799), pp. 205–206.

27. E. De Witt and M. Ames, eds., *Tree-ring chronology of eastern North America* (Tucson, 1978).

28. Orville Prescott, Jr., *A dissertation on the natural history and medicinal effects of the Secale cornutum, or ergot* (Boston, 1813).

29. M. F. Morgan, *The soils of Connecticut* (New Haven, 1930).

30. A. B. Appleby, "Diet in sixteenth century England: Sources, problems and possibilities," in *Health, medicine, and mortality in the sixteenth century,* ed. Charles Webster (Cambridge, 1979), pp. 97–116; Percy W. Bidwell and John I. Falconer, *History of agriculture in the northeastern United States, 1620–1860* (New York, 1941), pp. 13–14; and McMahon, "A comfortable subsistence," p. 52.

31. Samuel Wyllys Collection, Records of Trials for Witchcraft in Connecticut, Connecticut State Library, Hartford.

32. George L. Burr, ed., *Narratives of the witchcraft cases* (New York, 1968), pp. 28–29.

10. GREAT AWAKENING OR GREAT SICKENING?

1. For a sampling of views, see Darrett B. Rutman, comp., *The Great Awakening* (New York, 1970). Other recent studies include Harry S. Stout, "The Great Awakening in New England reconsidered: The New England clergy as a case study," *Journal of Social History* 8 (1974), pp. 21–41; Richard Warch, "The shepherd's tent: Education and enthusiasm in the Great Awakening," *American Quarterly* 30 (1978), pp. 177–198; Peter S. Onuf, "New Lights in New London: Group portrait of the separatists," *William and Mary Quarterly* 37 (1980), pp. 627–643.

2. Connecticut State Library, Hartford, Archives, Towns and Lands, ser. 2, vol. 7, p. 205, May 13, 1742; Billy G. Smith, "Death and life of a colonial immigrant city: A demographic analysis of Philadelphia," *Journal of Economic History* 37 (1977), pp. 863–889.

3. N. N. Reformatskii, *Dushevnoe Rasstroistvo pri Otravlenii Sporyn'ei (Bolezn' Zlaia Korcha)* (Moscow, 1893), pp. 96, 102, and 365–368.

4. N. N. Downing and W. Wygant, "Psychedelic experience and religious belief," in *Utopiates,* ed. R. Blum (New York, 1964), pp. 188–197; W. S. Pahnke, "LSD and religious experience," in *LSD, man and society,* ed. R. C. De Bold and R. C. Leaf (Middletown, Conn., 1967), p. 68.

5. Ralph Emerson, "Mycological relevance in the nineteen seventies," *Transactions of the British Mycological Society* 60 (1973), pp. 363–387.

6. Richard Schultes and Albert Hofmann, *The botany and chemistry of hallucinogens* (Springfield, Ill., 1980), p. 37.

7. Charles Chauncy, *The state of religion in New England* (Edinburgh, 1742), pp. 7–8 and 24; *Seasonable thoughts on the state of religion in New England* (Boston, 1743), pp. 169–170.

8. James Davenport, *Confessions and retractions* (Boston, 1944), p. 7.

9. Joshua Hempstead, *The diary of Joshua Hempstead* (New London, 1901), pp. 377–381.

10. Reverend Stephen Williams diary, Longmeadow, Massachusetts, Public Library.

11. Reverend Marston Cabot diary, New England Historical and Genealogical Society, Boston, Massachusetts.

12. *Sibley's Harvard graduates,* vol. 5 (Boston, 1937), pp. 317–322.

13. Eleazar Wheelock Papers, Connecticut State Library, Hartford, 742219.

14. Daniel Wadsworth diary, Connecticut Historical Society, Hartford;

Stephen Williams diary; letter from Reverend Elisha Williams to Solomon Williams, Connecticut Historical Society.

15. Hempstead, *Diary*, pp. 377–381, with some passages reconstructed with the help of Frances M. Caulkins, *History of New London, Connecticut* (New London, 1895), p. 450.

16. Nathan Cole diary, Connecticut Historical Society, Hartford. See also Michael J. Crawford, ed., "The spiritual travels of Nathan Cole," *William and Mary Quarterly* 23 (1976), pp. 89–126.

17. Reverend Jacob Eliot diary, Yale University, New Haven, Connecticut.

18. Daniel Wadsworth, *The diary of Daniel Wadsworth* (Hartford, 1894); the complete manuscript is in Connecticut Historical Society. See also Nathan Cole diary; Jonathan Edwards, ed., *Life and diary of David Brainerd* (Chicago, 1949).

19. "Religious excitement one hundred and odd years ago," *New England Historical and Genealogical Register* 15 (1861), pp. 23–24.

20. Sarah F. McMahon, "A comfortable subsistence: The changing composition of diet in rural New England, 1620–1840," *William and Mary Quarterly* 62 (January 1985), pp. 32 and 52.

21. U.S. Bureau of the Census, *Historical statistics of the United States* (Washington, D.C., 1975), vol. 2, p. 1196.

22. Jacob B. Felt, *Annals of Salem* (Boston, 1849), vol. 2, p. 200.

23. David Hall diary, Massachusetts Historical Society, Boston, Massachusetts.

24. Connecticut State Library, Hartford, Archives, Towns and Lands, ser. 1, vol. 7, pp. 203 and 205.

25. Charles E. Clark, *The eastern frontier* (New York, 1970), p. 139.

26. William Willis, ed., *Journals of the Rev. Thomas Smith and Rev. Samuel Deane* (Portland, Me., 1849), pp. 96–97.

27. *Public records of the Colony of Connecticut, 1636–1776* (Hartford, 1850–1890), vol. 8, pp. 376ff.

28. David Ludlum, *Early American winters, 1604–1820* (Boston, 1966), p. 15; Nathan Stone diary, Massachusetts Historical Society; William Wheeler diary, Connecticut State Library.

29. See note 19, above. No instrumental readings of temperature were kept in northern New England at this time.

30. Elizabeth D. Dodds, *Marriage to a difficult man* (Philadelphia, 1971), p. 91; Stephen Williams diary.

31. Thomas Prince, ed., *The Christian history* (Boston, 1744), p. 371.

32. For elevations see *Rand McNally 1980 commercial atlas and marketing guide* (Chicago, 1980). *Biographical sketches of the graduates of Yale College*, vol. 1 (New York, 1885), p. 462; Ellen D. Larned, *History of Windham County, Connecticut* (Worcester, Mass., 1974), vol. 1, p. 464.

33. Oliver Prescott, Jr., *A dissertation on the natural history and medicinal effects of the secale cornutum, or ergot* (Boston, 1813), p. 5.

34. Joseph Tracy, *The Great Awakening* (Boston, 1842), pp. 145, 153, and 163; James Walsh, "The Great Awakening in the First Congregational Church of Woodbury, Connecticut," *William and Mary Quarterly* 26 (1971),

pp. 543–562; Gerald Moran, "Conditions of religious conversion in the First Society of Norwich, Connecticut, 1718–1744," *Journal of Society History* 5 (Spring 1973), pp. 331–343; and Philip J. Greven, Jr., "Youth, maturity, and religious conversion: A note on the ages of converts in Andover, Massachusetts, 1711–1749," *Essex Institute Historical Collections* 108 (1972), pp. 119–134.

35. Leonard W. Labaree, "The conservative attitude toward the Great Awakening," *William and Mary Quarterly* 1 (1944), pp. 331–352.

36. John Winthrop, *Winthrop's journal: History of New England, 1630–1649*, vol. 1 (New York, 1908), pp. 97 and 240.

37. Robert Calef, *More wonders of the invisible world* (1700), in George L. Burr, *Narratives of the witchcraft cases, 1648–1706* (New York, 1914), pp. 308–314 and 322.

38. Thomas Robie, commonplace book, Massachusetts Historical Society.

39. C. C. Goen, ed., *The Great Awakening* (New Haven, 1972), p. 313.

40. William Douglass, *A summary, historical and political, of the first planting, progressive improvement, and present state of the British settlements in America* (Boston, 1750), vol. 1, pp. 100, 448–450.

41. George Whitefield, *Journals, 1737–1741* (Gainesville, Fla., 1969).

42. Francis G. Wallet, ed., *The diary of Ebenezer Parkman, 1703–1782* (Worcester, 1974).

43. Benjamin Lord, *God glorified in His works of providence and grace* (Boston, 1744).

44. Felt, *Annals of Salem*, vol. 2, p. 425; diary of Edward Holyoke, Essex Institute, Salem, Massachusetts.

45. Connecticut State Library, Hartford, Archives, Town, and Lands, ser. 1, vol. 8, pp. 86–87.

46. Hempstead, *Diary*.

47. William Currie, *An historical account of the climates and diseases of the United States of America* (n.p., 1792).

48. Wesley M. Gewehr, *The Great Awakening in Virginia, 1740–1790* (Durham, N.C., 1930), pp. 151–152, 167–169.

49. Percy W. Bidwell and John I. Falconer, *History of agriculture in the northern United States, 1620–1860* (New York, 1941), pp. 95–96; Chester M. Destler, "The gentleman farmer and the new agriculture: Jeremiah Wadsworth," in *Farming in the new nation*, ed. D. P. Kelsey (Washington, D.C., 1972), pp. 141–145.

50. Holyoke papers, Essex Institute, Salem, Massachusetts.

51. Ebenezer Gay papers, Suffield Public Library, Suffield, Connecticut; David Avery diary, Connecticut Historical Society, Hartford; Vine Utley, *History of a mortal epidemic* (n.p., 1813).

52. Peter Cartwright, *Autobiography of Peter Cartwright* (New York, 1856), pp. 51–52. For a summary of symptoms, see Catherine C. Cleveland, *The Great Revival in the West, 1797–1805* (Chicago, 1916), pp. 85–125; and Charles A. Johnson, *The frontier camp meeting* (Dallas, 1955), pp. 57–65.

53. John B. Boles, *The Great Revival, 1787–1805* (Lexington, Ky., 1972),

p. 56; Richard McNemar, *The Kentucky revival* (Pittsfield, Mass., 1808), p. 21.

54. Harriette S. Arnow, *Flowering of the Cumberland* (New York, 1963), p. 261; F. Marian McNeill, *The Scots kitchen*, 2d ed. (London-Glasgow, 1963), p. 178; and Marjorie Plant, *The domestic life of Scotland* (Edinburgh, 1952), p. 100.

55. Boles, *Great Revival*, pp. 1–2; John C. Campbell, *The southern highlander and his homeland* (n.p., 1921), p. 68.

56. Harriette S. Arnow, *Seedtime in the Cumberland* (New York, 1960), p. 394.

57. *Rand McNally 1980 commercial atlas and marketing guide* (New York, 1980).

58. Boles, *Great Revival*, pp. 46–48; Cleveland, *Great Revival*, p. 41; McNemar, *Kentucky Revival*, p. 28.

59. Lewis C. Gray, *History of agriculture in the southern United States to 1860*, 2 vols. (Washington, D.C., 1932).

60. John Stearns, "Account of the Pulvis Parturiens," *Medical Repository* 11 (1808), pp. 308–309.

61. Ulysses P. Hedrick, *A history of agriculture in the state of New York* (New York, 1933), p. 338; Bidwell and Falconer, *History of agriculture*, pp. 97 and 240.

62. Whitney Cross, *The burned-over district* (New York, 1950), p. 8.

11. SOCIAL CONTROL OF MASS PSYCHOSIS

1. Louis F. Calmeil, *De la folie* (Paris, 1845), vol. 1, pp. 219 and 254–256.

2. Owsei Temkin, *The falling sickness* (Baltimore, 1971), pp. 138–146 and 193–195.

3. Claus B. Clasen, *Anabaptism: A social history, 1525–1618* (Ithaca, 1972), pp. 122–136; George H. Williams, *The radical reformation* (Philadelphia, 1962), p. 443.

4. Hugh Barbour, *The Quakers in Puritan England* (New Haven, 1964), pp. 70, 119–120, and 233–234; Keith Thomas, *Religion and the decline of magic* (New York, 1971), pp. 139–142; A. L. Morton, *The world of the Ranters* (London, 1970).

5. Francis Higginson, *A brief relation of the irreligion of the northern Quakers* (London, 1653), pp. 15 and 18; and Barbour, *Quakers in Puritan England*, pp. 118–119.

6. William C. Braithwaite, *The beginning of Quakerism*, 2d ed. (Cambridge, 1961), pp. 125 and 167.

7. Ibid., p. 158.

8. Temkin, *Falling sickness*, pp. 139–141; George Barger, *Ergot and ergotism* (London, 1931), pp. 65–68.

9. Calmeil, *De la folie*, vols. 1–2, passim.

10. Barbour, *Quakers in Puritan England*, pp. 70–71, 232–235 and 250–252; Higginson, *Brief relation*, n.p.

11. Quoted by Ronald A. Knox, *Enthusiasm* (New York, 1950), p. 150.

12. Barbour, *Quakers in Puritan England*, pp. 119–120.

13. George Rosen, "Enthusiasm," *Bulletin of the History of Medicine* 42 (1968), pp. 393–421; Margaret C. Jacob, *The Newtonians and the English Revolution, 1689–1720* (Ithaca, 1976); James R. Jacob, *Robert Boyle and the English Revolution* (New York, 1977), and *Henry Stubbes, radical protestantism, and the early Enlightenment* (Cambridge, 1983); Hillel Schwartz, *Knaves, fools, madmen, and that subtile effluvium* (Gainesville, Fla., 1978); Schwartz, *The French prophets* (Berkeley, 1980); Carolyn Merchant, *The death of nature* (New York, 1980); Brian Easlea, *Witch-hunting, magic, and the new philosophy* (Brighton, 1980); Michael Heyd, "The reaction to enthusiasm in the seventeenth century: Toward an integrative approach," *Journal of Modern History* 53 (1981), pp. 258–280.

14. Charles Bost, *Les prédicants protestants des Cevennes et du Bas Languedoc, 1684–1700*, 2 vols. (Paris, 1912); Phillipe Joutard, *Les Camisards* (Paris, 1976), pp. 17 and 69–76.

15. Thomas Short, *A general chronological history of the air, weather, seasons, meteors, etc.*, vol. 1 (London, 1749), pp. 382–383; Schwartz, *French prophets*, pp. 18–19; Thomas D. Atkinson, *Cambridge, described and illustrated* (London, 1897), p. 113; David H. Hosford, *Nottingham nobles and the north* (Hamden, Conn., 1976), p. 118; Ralph Thoresby, *The diary of Ralph Thoresby* (London, 1830), pp. 188–189; Barbour, *Quakers in Puritan England*, p. 101; John Harland, *Lancashire folklore* (London, 1882), pp. 99–101.

16. Robert Kreiser, *Miracles, convulsions, and ecclesiastical politics in early eighteenth century Paris* (Princeton, 1978), pp. 173–176 and 157–167.

17. John Wesley, *Journal*, ed. Nehemiah Curnock (London, 1938), vols. 2–4, passim; Robert Whytt, *Observations on the nature, cause, and cure of those disorders which have been commonly called nervous, hypochondriac, or hysteric*, 3d ed. (Edinburgh, 1967), pp. 212–213; James Cornish, "On a convulsive epidemic in Cornwall, caused by religious excitement," *Medical and Physical Journal* 31 (1814), pp. 373–379; W. McIlwaine, "On physical affections in connection with religion as illustrated by 'Ulster revivalism," *Journal of Mental Science* 6 (1859–1860), p. 439, and 7 (1860–1861), p. 59; Eugene L. Backman, *Religious dances in the Christian Church and in popular medicine* (London, 1952), pp. 310–311; Hugo Froderberg, "The preaching sickness in Smaland, Sweden, in the 1840s," *Sydvenska Medicin Historiska Sallskapet Arsskrift*, 1965, pp. 79–94; F. Sirois, *Epidemic Hysteria* (Copenhagen, 1974).

18. Linnda Caporael, "Ergotism: The Satan loosed in Salem?" *Science* 192 (1976), pp. 21–26.

19. *U.S. News and World Report*, March 21, 1988, p. 56.

Index

Abbott, Sarah, 115–116
Age: and ergotism, 12, 77–78, 117, 133; and alimentary toxic aleikiia, 15; and plague, 49–50, 52; and immune-deficiency symptoms, 60; and witchcraft accusations, 71, 77–78, 114, 117; and mortality, 94–96, 101, 102–103, 104, 108; and the Great Awakening, 133
Alimentary toxic aleikiia (ATA): in Russia, 6, 22, 110–111; seasonal fluctuations in, 15; symptoms of, 16; misdiagnosed, 17, 22; and throat distemper, 107, 109–112
Alpine areas, witchcraft accusations in, 72
Alsace-Lorraine, 99
Anabaptists, 149–150
Anglia, East, 55
Artois (France), 82
Aspergillus flavus, 18
Aspergillus fumigatus, 13
Augsburg (Germany), 74

Baltic states, 24, 33, 39, 74
Barlow, John, 117
Basel, 55
Baxter, Richard, 152
"Bewitchment": psychological explanations of, 114, 115; symptoms of, 115–117. *See also* Witchcraft accusations
Biraben, Jean-Noël, 91
Birth rate. *See* Fertility
Black Death (*1348–1350*), 47–52
Bordeaux, 84
Boston: throat distemper in, 107, 108–109; sickness in, 137
Boyer, Paul, 113–114

Brabant (Belgium), 54
Branch, Catherine, 117
Braudel, Fernand, 4
Breast-feeding, 15, 29, 52, 65, 94, 95, 117
Brittany, 83, 84

Cabot, Rev. Marston, 126
Calef, Robert, 136
Calvin, John, 150
Cambridge (England): population historians in, 3–4; sickness in, 56
Camisards, 152–153
Caporael, Linnda, 113, 153
Catholic Church, 151
Caulfield, Ernest, 107–109
Cereals: staple of European diet, 5–6, 83–84; preparation of, 6, 17, 49–50, 52, 117, 140; storage of, 6–7, 17, 50. *See also* Corn; Price of grain; Rye; Wheat
Chauncy, Rev. Charles, 125, 126
Cheyne, George, 94–95
China, mold poisoning in, 6
Class differences: and quality of cereals consumed, 6, 14, 19–20, 55–56, 63; and plague, 52; and ergotism, 64; and witchcraft accusations, 75, 79; and fertility, 94; and mortality, 94, 104
Claviceps, differences among strains, 38–39, 57, 65, 69, 117, 133, 156. *See also* Ergotism
Climatic variables. *See* Moisture; Temperature
Cogan, Thomas, 61
Colden, Cadwallader, 110
Cole, Nathan, 129

Coley-Smith, J. R., 64
Collins, Zaccheus, 118
Congregational Church, 133–136, 147
Connecticut: throat distemper in, 107,
    111–112; witchcraft accusations in,
    113, 117; Great Awakening in, 126,
    127–130, 131, 132–135, 138
Corn: in colonial New England, 42,
    110, 157; T-2 toxin in, 109, 111–112;
    in Early Modern Europe, 157
Creighton, Charles, 64, 76

Dauphiné (France): and the Great Fear
    of *1789*, 81, 82; religious revivals in,
    152
Davenport, Rev. James, 126, 130
Death rate. *See* Mortality
Deering, Charles, 92
Demography. *See* Population trends
Demos, John, 114
Dickinson, Jonathan, 110
Diphtheria, throat distemper diagnosed
    as, 107–109
Disease, infectious: and population
    trends, 5, 20, 36–38, 55, 90; in colo-
    nial New England, 107, 130
Doctors. *See* Physicians
Douglass, William, 108–110, 138
Drummond, Alexander, 80
Drummond, J. C., 76

Edwards, Rev. Jonathan, 132, 137–138
Elevation: and mold growth, 14, 18, 22,
    120, 132–133, 140; and immune-
    deficiency symptoms, 60; in New
    England, 120, 132–133, 134; and
    witchcraft accusations, 120, 150; in
    Kentucky, 140; in Europe, 150, 156;
    in Russia, 156
Eliot, Jacob, 129–130
England: molds in present-day, 18; er-
    gotism reported in, 19; ergot and fer-
    tility in, 61–67; molds and mortality
    in, 67–69; witchcraft accusations in,
    75–79; rye consumption in, 86; popu-
    lation trends in, 90; mortality trends
    in, 92–96; religious revivals in, 150–
    151, 152–153. *See also* geographic
    subdivisions
Ergotism: types of, 9; symptoms of, 9,
    11–12, 13, 76–77; and fertility, 9, 57,
    61–67, 156; in Russia, 9–12, 22–26,
    38–41, 42; misdiagnosed, 12–13, 22–
    23, 64–65, 133–136; and mortality,
    12, 13, 23–27, 59, 67–68; names for,
    13, 23, 65, 94–95; reporting of, 18,
    24, 64, 83, 84–87; prevention of, 23–
    24, 83, 87; in medieval Europe, 55–

58; and witchcraft accusations, 76–
    79; and Great Fear of *1789*, 82–87
Essex (England): plague in, 51; witch-
    craft accusations in, 75–79
Essex County (Massachusetts), witch-
    craft accusations in, 113, 115–116,
    117–122
Ethiopia, ergotism in, 117
Europe, northern: diet in, 6; fertility in,
    41; sickness in, 60; plant and human
    health in, 155–158
Europe, southern, 6, 41, 60
Ewen, C. L., 75, 77
Eyam (England), 49

Fairfield County (Connecticut), 113,
    117
Famine: and population trends, 4–5, 20,
    38, 55, 90; in Early Modern Europe,
    59–60; in colonial New England,
    118–119, 131–132; reduced in Mod-
    ern Europe and Russia, 158
Fertility: and population trends, 4, 26–
    28, 41–43, 57; and ergot, 9, 57, 61–
    67, 133, 156; seasonal fluctuations in,
    14; in England, 14, 42, 59, 61–67; in
    Russia, 26–28, 33–36, 38–39, 41–43,
    156; in colonial New England, 42,
    133, 157; and temperature, 57, 66–67;
    in Early Modern Europe, 59, 61–67;
    and prices of wheat and rye, 65–67;
    in Europe of *1750–1850*, 89–90, 92,
    95, 98, 101–102, 104
Finland: ergotism in, 29, 41; fertility in,
    38
Fitch, Rev. Jabez, 111
Food poisoning. *See* Alimentary toxic
    aleikiia; Ergotism
Fothergill, John, 60
France: ergotism reported in, 19; Great
    Fear of *1789* in, 81–87; population
    trends in, 90; mortality trends in, 96–
    103. *See also* geographic subdivi-
    sions
Fusarium toxins, 6, 14–19, 51, 94, 98,
    101–102. *See also* Alimentary toxic
    aleikiia (ATA)

Garrett, Clarke, 86
Geoffrey, Dr., 82–83
Germany: molds in present-day, 19; er-
    gotism in, 74, 88, 140, 153; interpre-
    tation of nervous disorders in, 149–
    150. *See also* geographic subdivi-
    sions
Gilman, Nicholas, 130
Gilpin, John, 151
Gottfried, Robert S., 55
Gottlieb, Jack, 113, 114, 117

Goubert, Jean-Pierre, 83
Grains. *See* Cereals; Corn; Price of grain; Rye; Wheat
Great Awakening, 123–141; and ergotism, 124, 132–136, 138–139, 140–141; clerical reaction to, 136, 141, 147. *See also* Religious revivals
Great Fear of *1789*, 81–87
Grenoble, 86

Hainault-Cambresis (France), 101
Hall, David, 131
Hallucinations: in colonial New England, 116, 118, 123, 124, 125, 137–138; unpleasant, 124, 145; pleasant, 124, 146; in Kentucky, 139–140
Hamilton, Pat B., 7
Hammond, Lawrence, 119
Harris, Walter, 65
Harrison, William, 79
Harvey, Joan, 77
Hempstead, Joshua, 111, 126, 128–129
Higginson, Francis, 150
Hillary, William, 60
Holme, Thomas, 151
Holyoke, Edward, 138
Hungary, molds in present-day, 19
Huntingdonshire (England), 77
Hutchinson, James, 96
Huxham, John, 60
Hysteria, "bewitchment" diagnosed as, 114. *See also* Panic

Immune-deficiency symptoms, 59–61, 96, 101; in colonial New England, 107, 112, 123
Industrialization, and population trends, 4, 88–90, 157
Ingersoll, Nathaniell and Hannah, 116
Interpretation of nervous disorders, 13, 20, 71, 80, 86–87, 133–136; distinguished from causes of disorders, 123; dominant screens of, 145–153
Ireland: mortality in, 41; plague in, 51; witchcraft accusations in, 73
Italy, 49, 53; molds in present-day, 18; mortality in, 41; witchcraft accusations in, 74; ergotism in, 86

James II (king of Ireland), 153
Jarvis, Basil, 18
Jenkins, James, 86–87
Johnstone, James, 60
Jorden, Edward, 64–65, 151
Josselin, Ralph, 51, 63

Kemp, Ursula, 80
Kentucky: Great Awakening in, 123, 141; "nervous fever" in, 138

Kerridge, Eric, 76, 79
Kittredge, G. L., 75
Kolosov, G. A., 12

Labrousse, Suzette, 86
Lee, Ronald, 4
Lefebvre, Georges, 82
Leicestershire (England), sickness in, 56
Lepecq, Louis, de la Clôture, 101–102
Le Pois, Charles, 151
Life expectancy: in Britain, 3, 69, 95–96; in Russia, 69
Linsell, C. A., 7
London: temperature and disease in, 37; plague victims in, 49; mortality in, 96
Lorcin, Marie-Thérèse, 55–56
Lord, Benjamin, 138
Louis XVI (king of France), 82
LSD: in ergot, 9, 116; effects of, 116, 124
Lyon, Henry, 126
Lyon (France), 55–56

MacDonald, Michael, 77
MacFarlane, Alan, 75–78
McKeown, Thomas, 91
Maine, 131–132
Maisonnais parish (France), 82
Malthus, Thomas, population theory of, 3
Manchester (England), 86
Marriage. *See* Nuptiality
Marshall, Charles, 151
Massachusetts: throat distemper in, 107, 108–109; witchcraft accusations in, 113, 115–116, 117–122; Great Awakening in, 126–127, 131, 132, 135, 138
Methodology, 19–21
Midelfort, Erik, 74
Midlands (England), 86, 92
Moisture: and mortality, 5, 28, 29, 93, 97–99, 101; and mold growth, 6–7, 13–14, 18, 22, 25–26, 51–52, 58, 93; in Russia, 28, 29, 156; and plague, 49, 50–51; in medieval Europe, 50, 51, 57–58; and immune-deficiency symptoms, 59–61; in Early Modern Europe, 59–61, 156–157; and witchcraft accusations, 72–74, 78, 79, 119–122; in Europe of *1750–1850*, 84, 86, 93, 97–99, 101; in colonial New England, 111–112, 119–122. *See also* Temperature
Mortality: and population trends, 4; and temperature, 5, 28–36, 37–39, 50–51, 67–68, 93, 97–99, 101; seasonal fluctuations in, 5, 15; and

Mortality (*continued*)
moisture, 5, 28, 29, 93, 97–99, 101;
and ergot, 12, 13, 23–27, 59, 67–68,
86; in Russia, 23, 26–36, 37–38, 156;
in medieval Europe, 48–52, 53; in
Early Modern Europe, 59–61; in Early
Modern England, 67–68; and prices of
wheat and rye, 68; and witchcraft accusations, 76; in Europe of *1750–
1850*, 86, 89–91, 92–104; and age,
94–96, 101, 102–103, 104, 108; in colonial New England, 108, 125
Moryson, Fynes, 61–63
Murashkinskii, N. E., 15
Muscovite Russia, 25, 39, 156, 157
Mycotoxicoses, 7, 19. *See also* Alimentary toxic aleikiia; Ergotism

Nantes (France), 85
Napier, Richard, 77
Nervous disorders: symptoms of, 9, 70,
125–130, 136–138, 139–140; in Russia, 22; in medieval Europe, 56–57; in
Early Modern Europe, 60, 65, 70–80,
149–153; in Europe of *1750–1850*,
82–83, 95, 153; in colonial New England, 118, 123, 125–130, 136; in
Kentucky, 123, 139–140; interpretation of, 123, 145–153; diagnosis of,
147
New England, colonial, 42, 157. *See
also geographic subdivisions*
New Hampshire: mortality in, 108;
corn consumption in, 110; throat distemper in, 107, 111, 112; Great
Awakening in, 130, 131, 132
New Jersey, throat distemper in, 110
New York, religious revivals in, 140–
141
Nissenbaum, Stephen, 113–114
Norfolk (England), "bewitchment" in,
77
Normandy: population trends in, 54,
55, 101–103; and Great Fear of *1789*,
86
Nuptiality: in Russia, 26–27, 156; in
Europe of *1750–1850*, 88–90
Nutrition, and population trends, 4. *See
also* Famine

Orenburg (Russia), 15
Oster (Russia), 22

Panic: Great Fear of *1789* as incident
of, 82, 84–87; in Early Modern En-
gland, 153
Paris, temperatures in, 97, 103
Parkman, Ebenezer, 138
Parris, Rev. Samuel, 114

Pemell, Robert, 65
Périgord (France), 86
Périgueux (France), 54
Perm province (Russia), 13, 23, 26
Perpignan (France), 101
Physicians: in Russia, 22; in Early
Modern Europe, 60–61, 64–65, 76–77,
79–80, 94, 151; "witches" as, 77–80;
in Europe of *1750–1850*, 82–84, 94–
95, 96, 101–102; in colonial New England, 108–109, 125, 137, 138, 141
Plague, 48–52
Poland, witchcraft accusations in, 74
Pool, William, 151
Population trends: Malthusian theory
of, 3; Cambridge theories of, 3–4; and
industrialization, 4, 89–90, 157; and
famine, 4–5, 20, 38, 55, 90; and moisture, 5, 58; and disease, 5, 20, 36–38,
55, 90; in Russia, 26–43, 156, 157;
and war, 38, 90; in medieval Europe,
47, 48, 53–55, 58; and temperature,
58, 92; in Early Modern Europe, 59,
61, 65–69, 155–158; in Europe of
1750–1850, 88; and potatoes, 91–92;
in China, 92; in colonial New England, 92
Potato consumption: and mortality,
24–25, 28, 36, 91–92, 96, 99–100,
103; in Russia, 24–25, 30, 36, 37; in
colonial New England, 112, 131; in
Early Modern Europe, 157–158
Price of grain: and fertility, 65–67; and
mortality, 68; and witchcraft accusations, 74–79; in colonial New England, 131; in modern Europe, 158.
See also Cereals; Rye; Wheat
Protestant Reformation, 150
Psychosis. *See* Nervous disorders
Putnam, Ann, 113
Putnam, John, 116

Quakers, 86–87, 147, 150–151, 152

Raulin, Joseph, 83
Red bread, 13, 118
Reformatskii, N. N., 9–12, 115, 124
Religious revivals, 118, 123; in Europe
of *1750–1850*, 86–87, 153; psychological explanation of, 130, 140; in colonial New England, 136–141; in Early
Modern Europe, 150, 151–153. *See
also* Great Awakening
Rhineland: sickness in, 56–57; witchcraft accusations in, 72; Protestant
activity in, 150
Rhône Valley (France): witchcraft accusations in, 72; and Great Fear of
1789, 81

Riley, James, 90–91
Riolan, Jean, 149
Robie, Thomas, 137
Roussillon (France), 86
Rule, Margaret, 136–137
Russia, 9–12, 22, 115, 124; ATA in, 15, 22; distribution of mold poisoning in, 22–26, 39–41; population trends in, 27–36; witchcraft accusations in, 74. *See also geographic subdivisions*
Rye: versus wheat in bread, 6, 55–56, 61–63, 92, 156; and ergot, 7; in Russia, 28, 30, 36, 37, 41; in Early Modern Europe, 61–64, 72–74, 78, 79, 86, 150; and Great Fear of *1789*, 84–87; in Europe of *1750–1850*, 92, 94–95, 96, 102, 103; in colonial New England, 131, 138–139, 140, 157. *See also* Cereals; Price of grain; Wheat

Sadler, John, 65
Saint Anthony's Fire. *See* Ergotism
Salem (Massachusetts): witchcraft accusations in, 113–122; religious revival in, 136–137, 138; "nervous fever" in, 138, 139
Sandor, Gabriella, 19
Scandinavia: witchcraft accusations in, 74; population trends in, 90; religious revivals in, 153
Scarlet fever, throat distemper diagnosed as, 107–109
Schofield, R. S., 89–90, 92, 96
Scot, Reginald, 77
Scotland: witchcraft accusations in, 72–73, 80; religious revivals in, 153
Seasonal fluctuations: in mortality, 5, 15, 36–37, 97, 103–104; in mold growth, 17–18, 22, 78, 156; in throat distemper, 107; in witchcraft accusations, 118
Selford (England), 86
Sewall, Samuel, 119
Short, Thomas, 60
Sims, James, 96
Smith, Thomas, 131–132
Soil, and mold growth, 14
Sologne (France), 85–86
South Africa, molds in, 18
Spain, witchcraft accusations in, 73
Spanos, N. K., 113, 114, 117
Swabia (Germany), 74
Switzerland: witchcraft accusations in, 74; interpretation of nervous disorders in, 150
Sydenham, Thomas, 60–65

Tatar republic, 15, 17, 25, 26, 29–30
Taube, Johann, 64, 83

Taxil, Jean, 151
Temperature: and mortality, 5, 28–36, 37–39, 50–51, 67–68, 93, 97–99, 101; and mold growth, 6, 13–14, 18, 22, 27, 39, 76, 93, 132, 156; in Russia, 28–36, 38–39, 156; and plague, 49, 50; in medieval Europe, 50, 51, 57–58; and fertility, 57, 66–67; and witchcraft accusations, 58, 72, 74, 76, 80, 119–122, 150; in Early Modern Europe, 66–67, 72–74, 149, 150, 151, 156–157; in Europe of *1750–1850*, 84, 87, 92, 93, 97–99, 101; in colonial New England, 111–112, 119–122, 132. *See also* Moisture
Thacher, Rev. Peter, 126, 130
Thomas, Keith, 150
Thomas, K. V., 75
Throat distemper, 107–112
Throckmorton, Robert, 77
Tissot, Simon-André, 22, 95
Tituba, 113
Tricothecenes. *See* Alimentary toxic aleikiia (ATA)
Turin (Italy), 86
Tuscany (Italy): plague victims in, 49; population trends in, 53
Twigg, Graham, 48, 51

Ukraine, 22, 23, 25, 26, 39
United States: grain moisture levels in midwestern, 7; molds in present-day, 19, 24; plague victims in western, 49. *See also geographic subdivisions*

Venner, Tobias, 63
Viatka province (Russia), 9–12, 23, 26, 124
Virginia, 138
Volcanic eruptions: and ergot growth, 14; and mortality, 94

Wadsworth, Rev. Daniel, 127
Walle, Etienne van der, 4
War: and population trends, 38, 90; reasons for, 154
Watkins, Susan, 4
Weather. *See* Moisture; Temperature
Weyer, Johann, 149
Wheat: versus rye in bread, 6, 55–56, 61–63, 92. *See also* Cereals; Price of grain; Rye
Wheeler, Mercy, 138
Wheelock, Eleazer, 130
Whitefield, Rev. George, 138
Wilbraham, A., 76
Williams, Elisha, 127–128
Williams, Rev. Stephen, 126–127, 131, 137, 138

Willis, Thomas, 60
Winthrop, John, Sr., 136
Witchcraft accusations: in medieval Europe, 57, 58; and temperature, 58, 72, 74, 76, 80, 119–122; in Early Modern Europe, 70–80; and moisture, 72–74, 78, 79; and mortality, 76; in Essex (England), 75–79; nonrandom distribution of, 71–74, 79, 113–114; and ergotism, 76–79, 113, 114, 115, 153; in colonial New England, 113–122; as interpretation of nervous disorders, 145, 147, 149, 150; and elevation, 120, 150. *See also* "Bewitchment"
Wood, G., 64
Wrigley, E. A., 89–90, 92, 96

Xenopsylla cheopis, 49; and moisture, 51

Zinin, N., 13
Zwingli, Ulrich, 150